THE NATURALIST IN SOUTH-EAST ENGLAND

In the same series

THE NATURALIST IN CENTRAL SOUTHERN
ENGLAND
(Hampshire, Berkshire, Wiltshire, Dorset and Somerset)
by Derrick Knowlton

THE NATURALIST IN DEVON
AND CORNWALL
by Roger Burrows

THE NATURALIST IN THE ISLE OF MAN
by Larch S. Garrad

THE NATURALIST IN LAKELAND
by Eric Hardy

THE NATURALIST IN MAJORCA
by James D. Parrack

THE NATURALIST IN WALES
by R. M. Lockley

Other books by S. A. Manning

BROADLAND NATURALIST
THE RIGHT WAY TO UNDERSTAND THE COUNTRYSIDE
THE LADYBIRD BOOK OF BUTTERFLIES, MOTHS
        AND OTHER INSECTS
SYSTEMATIC GUIDE TO FLOWERING PLANTS OF THE WORLD

For Children

BAKERS AND BREAD
TREES AND FORESTS
THE INSECT WORLD
THE WOODLAND WORLD

THE NATURALIST IN

South-East England

KENT, SURREY AND SUSSEX

S. A. MANNING, FLS

DAVID & CHARLES : NEWTON ABBOT

ISBN 0 7153 6109 0

Set in 12 on 13 point Bembo
and printed in Great Britain
by Latimer Trend & Company Ltd
for David & Charles (Holdings) Limited
South Devon House Newton Abbot Devon

Contents

List of Illustrations

7

CHAPTER ONE

South-East England

Boundaries—Physical features—The Downs—
The Weald—The London Basin—The coast

ONE MAY ARGUE about the extent of 'South-East England', but for the purposes of this book it comprises Kent, Surrey and Sussex. These three counties constitute a clearly defined natural area of some 5,000 square miles. This region is bounded in the north by the river Thames with its estuary and in the west by the Hampshire basin. Along its southern and eastern margins lie the English Channel and the Strait of Dover, narrow waters that long ago helped to make the peninsula of the South East 'the gateway of England' and which now bring the modern menace of oil pollution to our shores.

The chalk ridges of the Downs run eastward across the area and reach the sea, the South Downs at Beachy Head on the Sussex coast and the North Downs at the South Foreland on the Strait of Dover. These ridges form the bottom and the centre of the green shape (the '£' sign, as J. E. Lousley called it) representing chalk on the geological map of England.

Together with the Butser hills of Hampshire, the North and the South Downs enclose and form a chalk rim to the Weald, a well-wooded area some 100 miles long and 40 miles broad, the 'Sylva Anderida' of the Romans, the 'Andredsweald' of the Anglo-Saxons. North of the North Downs lies part of the London basin, while south of the South Downs alluvial marsh occurs along parts of the coast.

The Downs

The North Downs extend from Farnham in the west through Guildford, Wrotham and Wye to Folkestone, and the South Downs from Petersfield to Eastbourne. These downland chalk ridges are interrupted by gaps carved by rivers. The main ones in the North Downs are those of the Wey, the Mole, the Darent, the Medway and the Kentish Stour, while along the South Downs are the gaps of the Arun, the Adur, the Ouse and the Cuckmere.

Both the North and the South Downs exhibit striking escarpments that slope inwards towards the Weald. These steep slopes are gradually moving, the scarp of the North Downs northward and that of the South Downs southward. As a rule, the dip-slopes of the Downs fall away gently, but in the Hog's Back, west of Guildford, both slopes are steep and the chalk outcrop is narrow. Dry valleys are a feature of the escarpments and dip-slopes of both the North and the South Downs.

Thin soils cloak the steeper downland slopes, while the gentler northern slopes of the North Downs and southern slopes of the South Downs are covered, here and there, with patches of clay, sand or gravel. On level tops of the North Downs there are large areas of 'clay-with-flints', but the chalk of the South Downs often has just a thin capping of loam.

The crest of the Downs never reaches 900ft. In the South Downs, though, Ditchling Beacon attains 813ft, Chanctonbury 783ft and Firle Beacon 718ft. The crest of the North Downs is generally lower, but it exceeds 700ft for much of the central section between the Wey and the Medway.

Woodland still covers much of the North Downs, the dominant tree being beech, a truly lime-loving species. The beechwoods are places of beauty at all times of year. In winter the clean greys of beech trunks contrast vividly with the browns of fallen leaves, now warm copper, now dull reddish, according to the play of light, and the greens of mosses carpeting the wood-land floor. In spring the beechwood assumes a softer, more

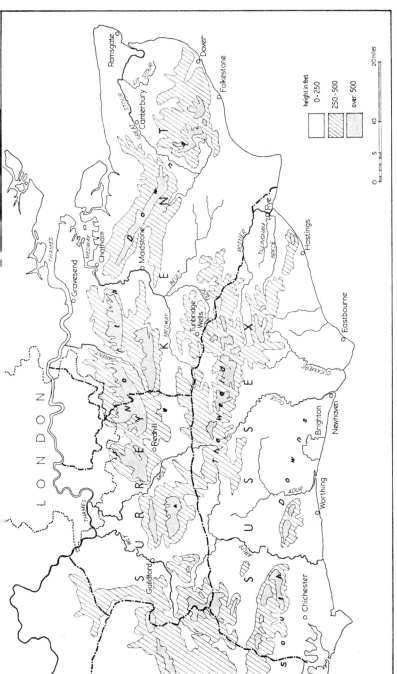

South-East England, showing main rivers and tributaries

delicate aspect as leaf-buds burst, revealing beautiful young leaves, pale green and fringed with silky hairs. In summer, it is true, the beechwood, close canopied with firmer, darker leaves, has a shady, almost austere, atmosphere that few green plants enjoy. But in autumn this gives way to a warmer scene as leaves become red-brown before falling once more before the winds and frosts of the year's end.

Small wonder, then, that Gilbert White considered the beech to be 'the most lovely of all forest trees'. He knew beeches and beechwoods in all seasons, in all moods. While admitting that beeches could be noble and indeed extraordinarily beautiful, A. C. Benson, that fascinating but now largely neglected writer, thought they were unfriendly. 'There is something almost repellently serene and virginal about them', he wrote, 'and I do not like the way they have of killing all the undergrowth about them.'

As we shall see in Chapter Four, certain shrubs and smaller plants do gain a foothold in even the most forbidding of the downland beechwoods. That these places should not be neglected by naturalists becomes obvious if one examines the leaves of beeches in open places where branches are able to sweep towards the ground. One soon realises that these leaves are the habitat of tiny leaf-mining insects, whose tunnelling creates curious patterns, and of minute gall-causing insects and mites whose galls consist of stumpy projections or of rolled or folded leaves. Each part of the beech has its features of interest. Even exposed roots may support mosses, the home of snail and woodlouse, or lichens, one of which is bright orange in light, open places and greenish or greyish in shadier situations.

Oak and birch, and locally hornbeam or sweet chestnut, thrive on the North Downs wherever there is an adequate depth of clay, sand or gravel over the chalk. In such places one may also find Scots pine, Corsican pine and European larch growing well. These local differences in the tree cover of the Downs help to create a variety of habitats and thus to increase the number of species making up the fauna and flora.

The South Downs between Worthing and the Hampshire border are densely wooded and here, as on the North Downs, beech is the most important tree. On the other hand, the Downs that lie east of Worthing support far less trees, there being only two really large woodlands in this thirty-mile stretch, namely Friston Forest, a downland water catchment area just north of the Eastbourne–Seaford road, and Stanmer Park, near Falmer.

The South Downs are, in fact, less varied and less extensive than the North Downs. For one thing, there is much less cover of superficial deposits above the chalk of the South Downs than on their northern counterpart. On the South Downs sheep farming was formerly almost universal, the excellent flavour of local mutton being attributed by some people to the presence of numerous snails in the close-cropped turf.

Important changes in land-use have since taken place there and sheep have disappeared from many parts. Following the outbreak of myxomatosis, which will be discussed later, rabbits, too, vanished from places where once they swarmed in large numbers. Thus changed farming methods and an introduced disease removed hundreds of thousands of animals whose grazing had helped to produce and maintain the treeless short turf so long associated with large expanses of downland. In many places sheep and rabbits have given way to plough and combine harvester and huge areas have been ploughed on slopes and crests alike. Arable crops and temporary leys have replaced long stretches of springy turf and even some of the steep valleys between the gentle southern slopes and parts of the steep northern scarps are fenced off and used for grazing cattle.

Much as many people regret the ploughing of the South Downs, one must admit that, with the disappearance of sheep and rabbits, it has, undoubtedly, saved much of our rolling downland from the rapid encroachment of shrubs and trees. This invasion by brambles, hawthorns and other elements of scrub, and later by trees, has already changed much of the natural beauty of the steep northern scarps and of areas disturbed by quarrying, as a glance at some of the illustrations in this book

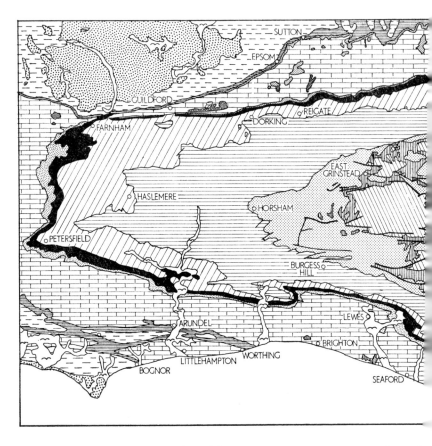

South-East England, showing geological features   Crown Copyright Geol

will make only too plain. In such places, the naturalist who
wanders off the thin turf of the track, with its outcrops of shiny
flints and white chalk, soon finds himself in a rank and grassy
wilderness whose coarseness does not favour some of the more
delicate chalk plants.

Fortunately, all is not lost and one may be thankful that
walkers, riders and cyclists have access to the South Downs Way,
Britain's first long-distance bridleway which runs for some eighty
miles between Eastbourne and the Hampshire border, mostly

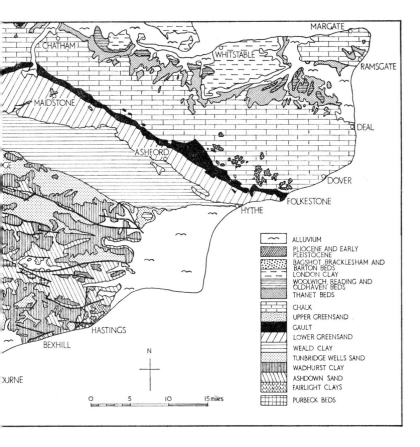

MARGATE
CHATHAM
WHITSTABLE
RAMSGATE
MAIDSTONE
DEAL
ASHFORD
GE
DOVER
FOLKESTONE
HYTHE
HASTINGS
BEXHILL
N
DURNE

	ALLUVIUM
	PLIOCENE AND EARLY PLEISTOCENE
	BAGSHOT, BRACKLESHAM AND BARTON BEDS
	LONDON CLAY
	WOOLWICH, READING AND OLDHAVEN BEDS
	THANET BEDS
	CHALK
	UPPER GREENSAND
	GAULT
	LOWER GREENSAND
	WEALD CLAY
	TUNBRIDGE WELLS SAND
	WADHURST CLAY
	ASHDOWN SAND
	FAIRLIGHT CLAYS
	PURBECK BEDS

O   5   10   15 miles

map Reproduced by permission of the Controller Her Majesty's Stationery Office

along the ridge of the South Downs. Along this ancient ridge-
way bridlepath two bird watchers have seen more than seventy
kinds of birds. Here, too, naturalists who are content to leave the
listing and recording to others may still enjoy the song of larks,
true creatures of open spaces and the air. Along this route there
are some very fine views southwards over the Downs, and also
fine and extensive views northwards over the Weald from
Windover Hill, Firle Beacon and Ditchling Beacon.

The Weald

Outstanding as these viewpoints of the South Downs are, they are not as high as places in the Weald. Leith Hill, the highest point in South-East England, reaches a height of 965ft on the Weald's encircling greensand ridges. This greensand country, a land of good drainage and abundant springs, bears soils that are often steep and 'sterile', though there are fertile, easily worked areas in valleys between the hills and in east and mid Kent.

Unlike those of the Downs, the greensand woods do not support many lime-loving trees or shrubs, though a few do thrive there. Their acid sands favour Scots pine and birches, trees quick to invade the heathy commons or 'charts' of the greensand country, other conifers and heather.

Heather can be a delightful plant and in the wilder, poorer parts of Britain has been a source of fuel, bedding, fodder and honey. Mature heather, however, can spread widely and casts a deep shade, ousting many light-demanding and less robust species. In dry weather it readily catches fire and areas blackened in this way are a sad but common sight, not only on the greensand but in many other parts of the South East. Heather, bracken and gorse roots may survive fires and eventually throw up new growth. But when heather that is over fifteen years old is burnt, its woody stems can generate sufficient heat to damage roots, and bracken may become established before the heather recovers. Serious heather fires also reduce, sometimes destroy, animal communities, and their prevention and control provides one good reason why local naturalists are often ready to act as volunteer wardens of places subject to heavy recreational use.

Yet the greensand and its immediate surroundings must not be regarded as country where only sombre conifers, purple bee-laden heather, and birches—whose joyous spring greens give way to autumn golds—may be seen. The Leith Hill district, to take but one example, is enlivened by a greater variety of flora and diversity of scene. Within a short distance of the tower on

Page 17 (*above*) Roe buck; normally a woodland species, the roe deer ventures into suburban gardens in search of privet, roses and other cultivated plants; (*below*) dark fallow buck; fallow deer can be troublesome when they start bark-stripping

Page 18
(*above*) Badger using special badger-gate; often regarded as a purely woodland animal, 'Brock' is even seen in busy built-up areas;
(*right*) fox scratching; many foxes now find food and safety in towns and suburbs

Leith Hill summit, from whose top thirteen counties can be seen, there are the renowned bluebells of Mosses Wood on the south slopes of the escarpment overlooking the Weald. Also in the neighbourhood are steep-sided valleys where one may trace the spring line where a narrow marshy belt supports a different type of vegetation at the junction of sandy beds with clay.

South of the escarpment, and less than a mile south-west of the summit of Leith Hill, are the woods of Leith Hill Place, with their rhododendrons and azaleas, but here we are already on the clay of a broad band of low-lying country within the Weald's greensand rim. To Cobbett, writing in 1823, this was 'the real weald, where the clay is bottomless'. Parts of this heavy but fertile clayland in the Vales of Kent and Sussex are still wooded, but much has been cleared and is now intensively cultivated and devoted to pastures, fields and orchards.

When one visits the Weald Clay vales today, it is not always easy to remember that this Low Weald was once under continuous, damp oak forest. Increasing demand for charcoal for the iron and glass industries resulted in widespread clearances in the fifteenth and sixteenth centuries, and later management methods gave rise to further clearances and the bringing in of conifers and other introduced species.

What are considered to be the examples closest to the original forest of the Weald Clay are the woods known as the Mens and the Cut, near Wisborough Green in Sussex. That excellent body, the Sussex Trust for Nature Conservation (formerly the Sussex Naturalists' Trust), realised the value of these woodlands and appealed for funds to help it purchase and manage them. At the time of writing the Trust has acquired the first parcels of the Mens woodland, comprising 116 acres of Idehurst Hurst and Crimbourne Wood. Here, on Weald Clay with seams of sandstone and 'Paludina' limestone, oak and beech have been undisturbed for a great many years, resulting in woodland containing trees of all ages and a wealth of dead wood on trees and on the ground. Hawthorn, blackthorn, hazel and crab apple grow under the oaks, and holly bushes under the beech. Mosses

B

are abundant and so are the birds, butterflies and other insects of old woodland.

Rising above the Weald Clay vales, and almost completely surrounded by them, the High Weald, a region sometimes called the Forest Ridges, extends from near Horsham to the sea near Hastings. This hilly plateau, the very heart of the Weald, reaches its highest point at Crowborough Beacon (792ft). Bearing a varied pattern of sands and clays, soft sandstone and silts, it is a land of woods and commons, orchards and hopfields, steep-sided sheltered 'ghylls' (ravines) and sandstone outcrops.

Though it still nourishes many woods, some of them re-planted since the 1939–45 war, the High Weald was once much more densely wooded. Ashdown Forest, stretching along the 700ft ridge between Crowborough and West Hoathly, is now largely heath and common, but 600 years ago, at the time of Edward III, it was a Royal Forest of 14,000 acres. In recent years it has been the scene of devastating fires, and suggestions have come from some quarters that, as H. L. Edlin, the distinguished forester, put it, 'the landscape would be improved if more of Ashdown were restored to its erstwhile wooded conditions'.

Woodland like St Leonard's Forest, near Horsham, has been renewed after much of it had been allowed to deteriorate into heathy wastes of bracken, bramble and heather. As the result of praiseworthy co-operation between the Forestry Commission and the Sussex Trust for Nature Conservation, it was possible in 1962 to establish three nature reserves in St Leonard's Forest, with the object of preserving small examples of the original forest. In one of these reserves, Sheepwash Ghyll, with its stream and waterfall, the canopy of oak, beech and alder maintains a damp micro-climate in which mosses and liverworts thrive. The second reserve, Mick's Cross, comprises about twelve acres of plateau woodland, largely of beech with an understorey of holly. The remaining reserve, Lily Beds, was established to protect a colony of lily-of-the-valley, whose deliciously scented white flowers are a well-known feature of the forest in spring, and whose survival there may well depend on dedicated folk who are prepared to spend time controlling bracken and heather.

The vital importance of co-operation between national and local organisations and the great value of help given by individual volunteers is again evident at Nap Wood, a fine example of central wealden oakwood situated on the A267 four miles south of Tunbridge Wells. Owned by the National Trust and leased to the Sussex Trust for Nature Conservation as a nature reserve, it is the habitat of badgers, foxes and fallow deer. Redstarts, wood warblers, great-spotted woodpeckers and nuthatches live there, as do mallard and moorhens which have nested on a small island in the middle of a pond. The flora includes several ferns, but it is the display of bluebells in May that excites the botanical specialist and country lover alike.

These examples are taken from Sussex because the major part of the High Weald falls within that county. But the naturalist who is prepared to explore this river-dissected, small-scale country on foot will also find much to interest him in the Kent and Surrey portions.

The London Basin

To the north of the Weald and the North Downs, and lying partly within the South-East region, is the London Basin, a funnel-shaped structural depression which received muddy and sandy sediments during Tertiary times. Formed from the muds, the London Clay has in places proved too stiff for cultivation and has remained as woodland in such places as Ashtead Woods in Surrey and Blean Woods near Canterbury.

Blean Woods National Nature Reserve extends to 164 acres, consisting almost entirely of oak woodland with a variety of coppice species, such as sweet chestnut, which are cut every few years. Certain very rare species of beetles and bugs are characteristic of these coppiced woodlands and, as a result of active coppice management, they returned to some parts of Blean Woods after prolonged absence.

Sandy and gravelly sediments from beneath the London Clay

give rise to some well-drained loamy soils in the fruit-growing areas of north Kent. Here the orchards with their acres of blossom are breath-takingly beautiful in spring, but one is brought speedily back to the stark realities of life and the whole business of earning a living if one pauses to discuss nature with a local fruit-farmer.

Regarded by many people as attractive birds, which they certainly are, bullfinches can become major pests in the eyes of a grower assessing damage done to buds (and sometimes open blossom) of cherries, gooseberries and plums, and later apples, pears and black currants. Brambles, whose fruit are prized by makers of bramble jelly and blackberry jam and whose blossoms are visited by many a colourful butterfly, are weeds to be eradicated in fruit-growing districts, for the bushes are a favourite nesting-place of bullfinches. Wood pigeons, whose voracious appetite is well known, will strip buds, including those of fruit trees, and so they are given even less encouragement here than in many places. Welcomed as pollinators, bees, and indeed other insects that perform this vital service, are seen as an essential part of fruit production.

Moving on to the coarse sterile sands of this part of the region, one finds extensive heaths, notably in the Bagshot district of Surrey. Nowadays great public use is made of many of these open spaces, and it is difficult to imagine that there was a time when George III could complain that the moors of the New Forest were more desolate than Bagshot Heath. But the observer will still find much of interest there, though one is forced to admit that in some cases only specialists have sufficient detailed knowledge to appreciate the real effects of human pressures on the smaller, more delicate, or lesser known plants and animals. The case of the very rare ant *Formica rufibarbis* illustrates the point. Reported to have been trodden out of existence in one locality in 1965, its extinction is said to be likely in another Surrey habitat, the only place on the mainland of Britain where it is now known to occur.

The coast

Nowhere has the destruction of habitats and species been more apparent than along the coast of the region. But, although it has been developed for much of its length, and is therefore artificial in many ways, the south-eastern seaboard retains a number of interesting parts and features.

There are fine chalk cliffs between Brighton and Eastbourne and between Folkestone and the Isle of Thanet, the legendary 'white cliffs of Dover' forming part of the latter stretch. Certain parts of these chalk cliffs still form the habitat of samphire and sea cabbage, yellow-flowered plants noticed on the chalk cliffs at Dover by the sixteenth-century botanist William Turner, the 'Father of English Botany'. Sadly, Shakespeare's Cliff, that famous landmark in the white chalk cliffs of Dover, is no longer the haunt of choughs, the red-legged crows or red-legged Jack Daws of the old writers, at least one of whom spoke of 'a war of extermination' having been ruthlessly waged against them. Gone, too, are the ravens, birds once heavily persecuted by local shepherds, whose cliff nesting-sites appear to have been used until about 1890. Fortunately, as we shall see, sea-birds and such other species as kestrels and jackdaws still live on the chalk cliffs and on the vertical cliffs of sandstone at Hastings and Pett Level.

Where softer rocks have formed bays, shingle, consisting mainly of flint pebbles, has collected. From the Kent coast the largest shingle structure in Britain, a complex series of shingle ridges or fulls, juts out into the English Channel. This great shingle foreland of Dungeness is, as we shall see, of considerable interest to geographers, botanists, ornithologists and entomologists, and it is a matter of great regret that it did not become one of our first National Nature Reserves and that it was allowed to be the site of a nuclear power station of the Central Electricity Generating Board.

On a smaller scale, but somewhat similar to Dungeness, is

Langney Point, or the Crumbles, near Eastbourne, a good spot for watching sea-birds and other species on passage. This large group of shingle ridges accumulated across a shallow bay and now forms a protective barrier on the seaward side of Pevensey Levels. Long since artificially drained, these 'Levels' comprise some 14 square miles of alluvial flats where deep drainage dykes are numerous. Here cattle graze and snipe, redshank and yellow wagtail breed. These three species also occur on other 'Levels', such as those of the Adur, Glynde and Rother. But being birds of rough grazings whose long coarse grasses effectively conceal their nests, they are not encouraged by ploughing, grassland improvement or the general tidying-up resulting from the more intensive agricultural use of 'Levels'. Increased drainage has also affected certain birds, particularly snipe which need boggy feeding places.

Like Pevensey Levels, Romney Marsh, another area of alluvial flats, was formerly a shallow bay. This silted up and was then artificially drained and reclaimed behind the vast protective barrier of the promontory of Dungeness. Romney Marsh has long been famed for its own breed of hardy sheep and they may still be seen grazing in the pastures of this level marshland, sometimes with lapwings and starlings feeding close to their feet. Cattle graze here, too, and good crops of potatoes, cereals, bulbs and greens are grown on the fertile, well-drained land. And, despite a widespread belief that the Marsh is treeless, one can find many a tree-lined lane. These, together with many deep, narrow ditches, the wider and shallower marsh fleets, and the slow, weedy streams, create additional habitats and add interest for the naturalist.

Turning to the western end of our regional coastline, there is Chichester harbour, mostly in Sussex but partly in Hampshire, a large, branched tidal inlet. This haunt of wildfowl and waders, where large numbers of brent geese may be seen in winter, is being used by rapidly increasing numbers of people, particularly for sailing. It is the subject of *The Chichester Harbour Study*, a publication of West Sussex County Council, the authority which has agreed that the main conservation areas should, torf he present at least, remain free from large-scale development.

Nearby Pagham harbour, a local nature reserve, is another wildfowl and wader refuge. Pagham harbour and its surroundings include extensive tidal marshland, mud-flats and waterways covered by salt and brackish water at full tides, the sea itself, Pagham Lagoon, reed-beds, water meadows, woods and fields. Thanks to this great variety of habitats, the area has attracted birds of more than 260 species and races and several mammals, while many kinds of plants grow there.

It had long been obvious that the noise and pressure of holiday-makers were severely disturbing, and even driving away, the breeding birds. But the declaration of the nature reserve in 1964 did not completely solve the problem overnight and naturalists owe much to the devotion and patience of those who watch over the little terns and other species during the breeding season.

Holidaymakers, speedboats, bait diggers and trespassers with guns have all helped to disturb wildlife and create problems for conservationists at the other end of the region's coast. However, like a plant that seems to thrive when trampled, or a willow that throws out straight new shoots when beheaded—which is just what pollarding amounts to—the Kent Trust for Nature Conservation has rallied support from individuals and organisations and has established nature reserves in a number of important areas, thereby deserving the support of all who appreciate the natural features, fauna and flora of coastal regions.

The Trust's acquisition of parts of Sandwich and Pegwell Bays, at the entrance to the Stour estuary, a large expanse of saltings, sand dunes, beach and tidal foreshore between Ramsgate and Deal, was rightly hailed by the late Garth Christian as one of the most important events in the history of the conservation movement in Britain. Here, in an area visited by waders, wildfowl, shore birds, and seals, are the habitats of sand dune and salt marsh plants and other species, including the rare carrot broomrape, a plant with pale yellow and purple flowers that is parasitic on wild carrot and less commonly on buck's-horn plantain and restharrow.

There was considerable disappointment when, against the representations of the Kent Trust and other bodies, it was decided to permit the development of a hoverport at Pegwell Bay. The limiting conditions imposed by the minister brought some relief, and it was encouraging to find the chairman of the Sandwich Bay Bird Observatory reporting that, in 1969, the hoverport's effect on the bird-life of the bay was not as discouraging as had been anticipated. Up to a thousand knot had been seen to congregate on the landing pads when they were not in use, but obviously the situation must be watched carefully.

Here, as in so many places, vigilance is essential. To achieve this full-time wardens will have to be employed in much larger numbers than at present, and educational campaigns will have to be mounted on a vast scale. Only in such ways will acts of vandalism be averted here, on the Kent coast, and elsewhere. And let no one be foolish enough to believe that vandals are confined to individuals, for acts of vandalism have been, and are being, perpetrated by government departments and by local authorities. The tipping of a town's rubbish on to the estuarine mud on the north side of the Stour, opposite the northern part of Sandwich Bay, for example, has affected the feeding grounds visited by waders at low tide.

That there is a crying need for more wardens and increased education in nature and environmental conservation is again evident as we move round the coast to the South Swale Local Nature Reserve, between Faversham Creek and Seasalter, where birds, including little terns, have been severely harassed by illegal shooting and intrusion by speedboats. Here, on the south bank of the Swale, by Graveney marshes, the beds of eel-grass (*Zostera*) attract brent geese, which feed on this marine flowering plant and winter in increasing numbers on the reserve. Other wildfowl come here, as do waders and seals, while the plants of this area of mudflats, salt marshes and rough, dyke-dissected grazing lands include fleshy-stemmed golden samphire and least lettuce, a close relative of the salad plant of gardens.

Further along the coast, as we travel towards the northern boundary of the region, are the extensive marshlands of the estuaries of the Medway and Thames, in parts of which industry, past and present, has made its mark. At High Halstow, in the central area of the North Kent marshes, is the wooded Northward Hill, a nature reserve and breeding site of numerous herons, birds that were once regarded as royal game, valuable assets for use in falconry and as items on the menus of great banquets, and in whose interest ravens were shot in Kent because they 'annoyed' the herons. The neighbouring estuarine marsh and wide foreshore attract large numbers of wildfowl and waders, the marshland being an important area for experiments and research into methods of increasing stocks of breeding wildfowl.

We end this survey of the South East, brief as it inevitably is, at the coast, an area of constant change, of erosion and accretion, of rise and fall of tide, of arrival and departure of man and migrant. Before we go on to consider the various types of wildlife of the region, let us remember that the many threats to the South East, with its rapidly expanding population, will not simply 'go away', and that continuous hard work, intelligent planning, and close co-operation will be needed from everyone in any way concerned with the use and development of its coast and countryside.

CHAPTER TWO

Mammals in the South East

Insectivores–Bats–Rabbits and hares–Rodents–
Cetaceans–Carnivores–Deer

DESPITE THE PRESSURE of its growing human population, the increase of traffic and the spread of building, the South East continues to offer much to interest the student of mammal life. This does not mean that naturalists can afford to be complacent or to disregard the large number of road casualties, now including such species as the dormouse and the mole, or to overlook the numerous creatures destroyed in heath and forest fires or trapped in discarded bottles. The extent of this last-named hazard to small mammals is not always appreciated, even by many active field naturalists. But the sad fact of the matter is that thirteen species have been identified among the victims of accidental 'bottling', common shrews and bank voles being the most frequent.

It has been suggested that most animals enter bottles out of curiosity or in search of food, though elderly individuals may crawl into these quiet and sheltered places to die. Small mammals seem to find no real difficulty in squeezing themselves through the holes of bottles. Escape is another matter, glass that is slippery and perhaps also wet often making this completely impossible.

Insectivores

The hedgehog, *Erinaceus europaeus*, is an example of the creatures suffering from modern hazards. The decline in its population, which has affected not only the South East but many other parts of England, is partly due to the members that are killed on the roads. As its natural reaction is to curl up on the approach of danger, trusting to the protection of its dense coat of spines, it has fared badly, though it is claimed that many suburban hedgehogs now make a rapid escape on the approach of a car. However, the fittest may survive and hedgehogs better able to cope with modern traffic may well evolve. Certainly 'urchins' can move fast, as I discovered for myself late one evening when a 'rat' that bolted from the churchyard in the centre of Shoreham-by-Sea proved to be a hedgehog.

Predatory Mammals in Britain, a code of practice for their management, which was first published in 1967 by the Council for Nature in association with the British Field Sports Society, the Fauna Preservation Society and the Game Research Association, states that, 'On a partridge beat nest losses to hedgehogs can be a serious factor,' but emphasises that, 'Elsewhere there is no need for control.' However, the ancient practice of persecuting the species continues in certain areas, and it is often convenient to ignore the fact that many partridge nests are destroyed by other predators and by mowing machines (not to mention the effects of weather conditions on game chicks).

Pesticides used against slugs, caterpillars and other natural foods of the hedgehog may have played their part in the animal's decline, the poison residues having accumulated in the vital fat that is stored in its body before hibernation. Cattle grids, garden ponds and sunken woodland water tanks take a toll of hedgehogs, while another significant factor is predation by badgers and foxes, which leave evidence in the form of skins turned inside out.

It is not perhaps surprising that many people seem to know little of hedgehogs and their ways as these creatures are largely active at night. Often regarded solely as inhabitants of hedgerows, copses, farmland and the edges of woods (mainly those of deciduous trees), they sometimes live at a lower density in open country than in parks, gardens, churchyards and cemeteries of suburban areas where dogs will often detect them as they shelter in tussocks in rough grassy places or in fallen leaves among shrubs and brambles.

The hedgehog's breeding season starts soon after the end of hibernation, in late March or early April. Selecting a site in a large grass tussock, under a dead tree or a woodpile, or in a dis-used or little used shed or outbuilding, the female constructs a large breeding nest, lining it with grass, moss and leaves. After a gestation period of 30–35 days she gives birth to a litter of from two to seven blind and hairless young whose skin is grey on the back and pink elsewhere.

The female may produce a second litter, but by about October or November she, like the rest of her species, will probably pre-pare for hibernation. Smaller and more compact than the breed-ing nest, the hibernating nest is hidden in a hedge bottom or compost heap, under a tree stump or log, beneath low tangled brambles or in an outbuilding. I have come across a record of a hedgehog that was not content simply to build a winter nest in a disused outhouse, but which appears to have moved several large onions some twenty yards and placed them at regular intervals near its heap of leaves and small sticks.

A hedgehog does not automatically fall into a winter sleep. If it is internally ready physiologically when the surrounding temperature reaches the critical level of 17° to 15° C it takes to its nest and rolls up tightly. Hedgehogs do not store food in the nest or elsewhere and many of those awakened in severe winters by sudden sharp spells of frost are believed to die of hunger. In milder winters hedgehogs often prove to be erratic hibernators, uneasy sleepers whose life span of some three to ten years con-sists of annual cycles that comprise four or five months of dor-mancy and seven or eight months of activity.

Although it is a true creature of the night, the hedgehog is sometimes seen foraging at dawn and dusk. Grunting and snort-ing, it searches for food, relying mainly on a sharp sense of smell, its eyesight not being particularly keen. Few serious studies have been made on this animal's food, but Konrad Herter, the distin-guished authority on the species, has recorded that in captivity one hedgehog ate 1,880g of mealworms in ten days, increasing its body weight by 466g, and 1,462g of sparrows in the following ten days, reducing its weight by 63½g.

We know that our south-eastern hedgehogs eat both animal and vegetable foods. They search orchards for windfall apples and pears, eating the fruit and turning them over to get at slugs that may have collected on the under side. Some tough-skinned species of slugs may be unpalatable to them, but they often manage to deal with slimy kinds by rolling them in the dust with their chins. The ground is searched for earthworms, some gardeners considering the hedgehog's ability to keep lawns free from worm-casts more effective than its efforts as a slug-hunter.

Hedgehogs are particularly fond of insects, their larvae and pupae. On being released from the pit beneath a cattle grid, one hungry individual spent several minutes snatching bees from a swarm that had formed under a hive (one hedgehog was stung by fifty-two bees without suffering any apparent ill effect!). Carrion, frogs and toads are also eaten by hedgehogs, while Jean Kenward, whose poem 'Hedgehog' graced the pages of *The Countryman* a few years ago, sketched a delightful picture of one which, having enjoyed its fill of raspberries, waddled 'belly-tight under the hedge for his repose'. Great drinkers, hedgehogs enjoy fresh water and regularly visit gardens where milk (or bread and milk) is provided.

Given the chance, 'urchins' do feed on the eggs of such ground-nesting birds as partridges, quails and corncrakes, but there are good reasons for believing that hen and pheasant eggs are not eaten unless they are already cracked. Certainly experiments have shown that hedgehogs could not open eggs of domestic hens and pigeons, while another observer found that these animals ate common gull eggs but only after they had been opened for them. Stories of hedgehogs taking eggs should always be treated with caution and naturalists should lose no opportunity of recording precise observations on this subject.

More first-hand accounts are also needed of the unexplained practice of 'self-anointing', a curious activity in which hedgehogs have been seen to engage. Having first sniffed at an object or a substance (usually one with a strong taste or smell), the hedgehog licks it all over or chews it thoroughly without actually swallowing or eating it. This action stimulates the flow of copious foamy saliva which is spread on the bristles, the animal turning its head and using its tongue for this purpose. Among several things and materials mentioned in connection with this salivation behaviour are glue used on the spines of books, soap, certain flowers and cigarette ends.

An especially interesting display of 'self-anointing' occurred in the case of a young hedgehog which had been deserted by the mother before its eyes opened and was taken care of by the observer who reported on its behaviour. Each day, on being released for an hour, the animal chewed dog excrement until a brown foam exuded from its mouth. After placing the foam on those parts of its body within reach, it set off on its exploration of the garden. This case certainly seems to support the suggestion that hedgehogs use the saliva to disguise their own scent and thus protect themselves from potential enemies.

Another strange type of hedgehog behaviour observed in the region is that of running (walking, according to some people) in circles, an apparently aimless activity that is kept up for some time. Since it was first described in 1967 by Mr J. S. Boys Smith, Master of St John's College, Cambridge, this curious phenomenon has been reported on a number of occasions, but naturalists should not neglect to record details of any future occurrences.

A few years ago the editor of *The Sussex Mammal Report* expressed the hope that naturalists would make a special attempt to keep records of hedgehogs, pointing out the importance of investigations into litter size, breeding behaviour, feeding habits and causes of death. It is, indeed, to be hoped that more and more people will return to the inquiring spirit of the old naturalists, testing theories by patient observation and, whenever necessary, relegating that which belongs there to the realms of folk-lore or fantasy.

Many people will, for example, have heard that hedgehogs are proverbially lousy, and a simple investigation has shown how large an animal population one can support. After bloated ticks were removed from its ears, the hedgehog was placed inside a polythene bag, its head protruding. Air was then pumped through a chloroform bottle into the bag and fleas immediately started to drop off the animal. Eventually 31 large ticks and 584 fleas were counted, but there were still hundreds of tiny ticks embedded in the creature's limbs and the skin of its underside. Careful examination of the animal's nest will often reveal the presence of the larval form of the hedgehog flea.

Fleas are also found in most nests of the mole *Talpa europaea*. This velvety-furred species builds its nest of dry dead leaves or withered grasses, placing it in its underground runway system under a 'fortress', a large mole-hill about a foot high and a yard across. These 'fortresses' should not be confused with the smaller, ordinary mole-hills, the 'nunky-tumps' of Norfolk, which are simply mounds of surplus soil pushed up as the mole excavates its deeper burrows. Often noticed on arable land in summer, the animal's shallow burrows appear as ridges on the surface of the ground. These surface workings often bring the presence of moles to the notice of predators. But that which may adversely affect one creature may benefit another, and robins, blackbirds, thrushes and fieldfares have learned to follow the surface workings of moles, taking any suitable food as it is exposed on the freshly turned soil.

The mole itself feeds mainly on earthworms and the larvae of beetles and flies, less important items of its diet being slugs, centipedes, millipedes and vegetable matter. A mole brought indoors by a cat (which also carried shrews in unharmed) lost all fear when offered worms, scraping them free of dirt before devouring them. During periods of drought, when earthworms and grubs are not easy to find, moles move to moister areas such as the banks of streams. One very dry summer they could be watched searching for food in such places, sometimes in the open, sometimes just below a thin mat of grass, their every movement visible to observer and predator alike. So thoroughly were the remaining remnants of damp ground scoured for worms and grubs that the ridges of the moles' surface workings were but a few inches apart and tunnels crossed and re-crossed themselves.

During their life span of three years (or less) moles travel much not only in search of food but when, in wet weather, they move to higher ground from water-meadows, marsh lands and other low-lying areas, land they will return to before it has had time to dry out after flooding. During the last few years moles have moved into certain dry heathland areas of the region following building development and the introduction of large quantities of loam and leaf mould for lawn and garden-making. Commonly recorded from farmland, woodland margins and churchyards, moles are also inhabitants of woodland, though coniferous woods do not seem to attract them.

Mole hills can be a nuisance in pastures and burrowing moles occasionally play havoc with young seedling trees in forest nurseries. In West Sussex moles have invaded glasshouses and are regarded as serious pests. Elsewhere they are considered to do great damage in gardens and forest rides.

The mole's natural predators do not seem to make great inroads into a population as much as two-thirds of which may die each year. Tawny owls take moles, especially from about May to August when the young animals are searching for unoccupied areas. Badgers will devour moles and at Kingley Vale they have fallen victim to foxes, their skins being left turned completely inside out. Man is the mole's most deadly potential enemy and there is no doubt that in some areas his attempts to control its numbers are successful. There is, however, a considerable measure of doubt as to whether the use of strychnine should ever be permitted. Despite the fact that this extremely dangerous substance is available only under permit and that it may only be used with earthworm bait and laid in deep mole runs, one is left wondering whether harm is being done to other wildlife and even to domestic animals and is being attributed to other causes.

Page 35 Beech in Slindon Park, Sussex; lichens grow on the tree trunks, while cushions of moss occupy the woodland floor

Page 36   Looking east along the South Downs from the summit of Ditchling Beacon (813ft), Sussex

The common shrew *Sorex araneus* is numerous in woods, copses and hedgerows in some areas, and the minute pygmy shrew *Sorex minutus* appears to be abundant, low undergrowth in gardens and along banks and old walls being among its haunts in the region. On the other hand, the water shrew *Neomys fodiens*, which may be found in woodland and other places far from water, seems to have suffered from the cleaning out and deepening of streams and ponds.

Much remains to be discovered about the status and habits of shrews in the South East and elsewhere. Living as they do in tunnels below ground and in runways among dense vegetation, these small mouse-like animals are seldom seen, though their high-pitched squeaking gives their positions away to experienced observers. Trapping apart, there are nevertheless a number of ways in which much can be gleaned about shrews. Counts can be made of those brought indoors by cats, which rarely eat shrews and do not always kill them, being sometimes inclined to carry them in alive merely to play with them until such time as they escape or are removed to safety by some kind-hearted person.

The pellets regurgitated by owls as they rid themselves of the fur and bones of their prey are another valuable source of information concerning shrews and other small mammals of a district. At Sedlescombe, Sussex, 29 barn owl pellets contained the remains of seven species, including 15 skulls of the common shrew, while another 85 pellets there included 39 shrew skulls, only two being those of pygmy shrews. Examination of pellets at Lydd showed that long-eared owls took a few shrews, and the same method established that tawny owls also preyed on these creatures on rough grassy parts of the Downs and in woods.

Other known predators of shrews include pheasants, not always suspected of eating animals of this type; they have been seen to eat them in our region, one Sussex pheasant having played with its victim, a pygmy shrew, before killing and swallowing it whole. Kestrels have been seen to take common shrews along hedgerows and on marshes, while the remains of

C

one of these small mammals were found in pellets ejected by white storks that formed part of a recent small invasion into south and east England. The part played by stoats and weasels in the control of shrew populations is often discussed, sometimes somewhat vaguely in the absence of serious studies of the food of these two carnivores. There is no doubt, though, that weasels do take shrews. One marshland weasel was so concerned about the shrew in its mouth that it almost ran into the observer's foot!

Peter Crowcroft, in his study *The life of the shrew* (1957), emphasised that the feeding of tawny owls was a very important factor in the reduction of the numbers of shrews, while starvation may well be the cause of death of many shrews in late summer and autumn. Due to these and other causes few shrews attain their full span which, at best, extends only through two summers.

During their short lives shrews are extremely active. Common and water shrews are very energetic burrowers which seize stones and other obstacles in their jaws and hastily remove them from the scene of their labours. Common and pygmy shrews are capable climbers, but water shrews seem to lack such ability. Common shrews will take to water in an emergency, but 'shrimp-mice' (as water shrews are called by the water-cress growers of Surrey) are naturally graceful and buoyant swimmers whose food is found in the water and on the land.

Solitary by nature, shrews will engage in noisy screaming fights ('sham fights' might often better describe them), driving other individuals away from their territory. This aggressive behaviour appears to cease only during the short periods of time when females are in breeding condition. It has been suggested that it helps to separate shrews whose feeding areas may overlap.

Being short-sighted and lacking a really keen sense of smell, shrews use their long whiskers to find food, sweeping them about swiftly and thoroughly. Observation of captive common shrews has revealed that, when several species of invertebrates are present, they exercise discrimination in their choice of food. One individual ate strawberry snails, beetles, woodlice (preferring *Philoscia muscorum* to the commoner *Porcellio scaber*, a choice noticed by several observers), and slugs, including the field slug and even the large and sticky black slug. Grasshoppers are also eaten and so are earthworms. Shrews are not exclusively carnivorous, however; they eat vegetable matter in small, but nevertheless essential, quantities.

Bats

Time may have dispelled the superstitions that once surrounded shrews, but one is forced to admit that the presence of bats still occasionally gives rise to all sorts of groundless fears. In recent years the Earl of Cranbrook has gone to considerable trouble to test the truth of the hoary superstition concerning the supposed irresistible attraction of women's hair for bats. Contrary to the gloomy predictions of those who have blindly passed on the ancient belief, he did not find it necessary to use scissors to separate bats from his assistants' hair. When put into the young women's hair, the bats remained completely unentangled, scrambled up over the top of the head and flew away. This result was obtained with four different species of bat and two eighteen-year-old blondes and also with an older brunette and a pipistrelle.

What a pity it is that more people do not make such attempts to seek out the facts and then act in a rational and humane manner. As it is, the presence of bats in buildings sometimes leads to what can only be called atrocities being committed. As recently as 1969 eighty pipistrelles were starved to death when the crack above a church door where these beneficial and harmless creatures were entering and leaving was blocked up.

In addition to churches, the pipistrelle *Pipistrellus pipistrellus* inhabits houses (both new and old), outhouses, trees and rock crevices. It is the commonest and smallest of the British bats, most of which have been recorded in the South East at some time or other. Although the pipistrelle is said to hibernate normally from the end of October to early March, it does fly during mild spells in winter and has even been observed in day-light in bitter weather in January and February. Remains of members of five different orders of insects have been found in its dung and so it must stand a good chance of finding a certain amount of food in winter when some, though fewer, insects appear in the open.

In June and July, when the female is busy feeding its single young, insects are, of course, far more plentiful. During these and other spring and summer months pipistrelles feed on gnats and other insects, mainly on the wing. They alight to drink or do so on the wing, hovering over the water with wings up-raised. Bird baths are visited for this purpose, pipistrelles some-times being seen to go right into the water and then to paddle across the bath to the other side.

Their rapid jerking flight occurs as much as forty feet up, but may take place as low as six feet above the ground, a fact which helps to explain why pipistrelles are numbered among road casualties. When they come really close to the ground, these small bats may be seized by cats. One such capture produced some valuable information. Ringed as an adult male in 1958, this particular pipistrelle was caught by a cat in 1964 when it was found to have moved 43 miles from the place of ringing, giving what is believed to be the longest recorded movement for the species in the British Isles. Remains of pipistrelles have been found in barn-owl pellets, but the use of stuffed owls in attempts to scare bats at such places as Horsham and Chichester does not appear to have met with much success.

The fact is that bats do not rely to any great extent on their eyes for identifying or avoiding objects. They navigate and also find and capture their prey by echo-location, giving out ultra-sonic impulses at frequencies far above the compass of the human ear. These impulses (or 'sonar') can be picked up by certain electronic machines which reproduce them at a much lower frequency, rendering them audible to the human ear.

Like the pipistrelle, the long-eared bat *Plecotus auritus* is a common road casualty and a species that suffers from the attacks of cats when flying low. Usually associated with trees or hedgerows, the long-eared bat eats moths, butterflies and certain other insects. Six of these bats made an attractive sight as they fed on moth larvae which were hanging on their threads from oak foliage. A light hibernator, the long-eared bat retires to buildings in October or later in the year, often coming out to drink at midday in winter. In some parts of the country this bat has taken to roosting in special boxes placed high on trees. Somewhat similar in design to a standard bird-box, the 'bat box' has its entrance as a slit in the bottom at the back and saw cuts are placed around the inside to give the bats footholds from which to hang.

The long-eared bat is given to flying near the ground or running quite fast along it (it can also run up a wall!), but the noctule *Nyctalus noctula* is often seen flying high in a rather swift-like manner. A species whose roosts occur in buildings and trees, the noctule has been known to take over an old green woodpecker hole in a tree-trunk after starlings had left it. It may be observed hawking with swallows over lakes and reservoirs, while the townsman should not be surprised if he sees it hawking round street lamps, even in drizzle. Before they are finally covered with soil, rubbish tips are a favourite haunt of noctules and certain other bats. At one such place the Earl of Cranbrook studied noctules by marking and recapturing those that came to feed on crickets.

The serotine *Eptesicus serotinus* also feeds over rubbish dumps and, like the noctule, will concentrate above places where cock-chafers are flying. Serotines inhabit buildings and hollow trees. Some of their colonies are well established, one Sussex colony having existed in the roof of a bakery from at least 1909 until it was closed in 1964. Serotines are seriously affected by the cold and, not surprisingly, their winter sleep is long and deep.

Sometimes called the forest bat in West Sussex, where a new site for this species was found a few years ago, Natterer's bat *Myotis nattereri* is a woodland species. It picks spiders and insects from leaves, but takes other food in flight, often after deliberately flying against foliage to disturb any insects resting or sheltering there.

Much remains to be discovered about this and other species of bats. Recent observations of interest include Major W. W. A. Phillips's discovery in Sussex of two colonies of the mouse-eared bat *Myotis myotis*, only the second record for this species in England since 1888, and the same worker's rescue of the first Sussex specimen of the grey long-eared bat *Plecotus austriacus* from a dustbin.

Rabbits and hares

Unlike most species of bats, rabbits and hares are readily recognised by townsman and countryman alike. Severely reduced in 1954–5 by myxomatosis, the disease spread by virus-carrying fleas, populations of the rabbit *Oryctolagus cuniculus* are re-establishing themselves in parts of the region. For, though still present, the disease does not seem to be affecting rabbits as seriously as it once did.

Rightly regarded as major pests of farming and forestry, because of their liking for growing crops and young trees, rabbits (together with sheep or cattle in some areas) have, however, played an important part in checking the development and spread of thorn scrub on downlands and other grasslands.

Until the late fifties of the present century, the chalk heath vegetation at Lullington Heath National Nature Reserve, which lies on the South Downs between Litlington and Jevington in East Sussex, was almost completely open. There were only small patches of dwarf gorse nibbled short by rabbits which, assisted by sheep, kept scrub down. With the decline in sheep grazing and the destruction of rabbits by myxomatosis, gorse bushes have grown rapidly in height and size at the expense of the chalk heath until the Nature Conservancy has considered it necessary to control their spread by mowing.

Similar problems have arisen at Kingley Vale, a National Nature Reserve in West Sussex that is famous for its yew trees.

Here under-grazing, due to the decline of rabbits through myxomatosis, has permitted scrub and woodland to develop rapidly on open grassland. Machinery is now being used to control scrub and mow grassland to maintain conditions suitable for the germination and establishment of the yew.

Such stories could be told of many other areas in our region and elsewhere. But it should be remembered that there are numerous places where the destruction of rabbits is regarded as important, so that less crops are lost to a pest and more young trees are able to mature.

The natural predators of rabbits include foxes, stoats, cats and dogs, weasels, badgers, crows and certain birds of prey. It was not surprising therefore that several of these creatures were seen to turn to other prey and to exhibit what some people regarded as changed habits after myxomatosis. With the rabbit population decimated, the field vole became the principal prey of the fox in South-East England, brown rats also being eaten in large numbers.

There has been much discussion as to the extent to which foxes were affected by the widespread destruction of rabbits by myxomatosis. Brian Vesey-Fitzgerald claims that 'even at the height of the rabbit era, voles and rats formed the main food of foxes'. But he admitted that a great many rabbits were taken by foxes before myxomatosis.

As has been stated authoritatively, until the outbreak of myxomatosis rabbit was the staple food of the stoat in the British Isles, though this small carnivore has long been known to take other small mammals, birds and eggs, too. After myxomatosis the stoat population fell to a low level in the region and stoats, often regarded as earth-bound creatures, were commonly observed climbing trees in their search for food. In Sussex a stoat killed and ate four young grey squirrels in a drey 22ft above ground, while in July a stoat managed to drag a young starling from a nest hole 25ft up an ash tree and to clamber down the tree trunk sideways with the bird in its mouth. Now, with rabbits on the increase in some areas, stoats seem to be re-establishing themselves in some numbers again.

The disease caused by the virus *Myxomatosis cuniculi* appears to be specific to rabbits and does not infect any other animals. Certainly the brown hare *Lepus europaeus* has not been affected by it. Indeed, within a few years of so many rabbits being eliminated by the disease, marked increases were noticed in the hare population of the region, though a slow decline set in later.

It soon becomes obvious to anyone driving through the South East that many hares are killed on the roads. Elsewhere they are sometimes found dead without apparent injury. Some people are inclined to believe that the reduction in the hare population in some places may be partly due to the increase in the number of foxes which will, of course, take hares, particularly the leverets. Again one cannot ignore the potential threat of a rising stoat population to young hares.

Whatever may prove to be the true causes of the apparent fluctuations in the hare population, we do know that this animal's distribution can be strangely patchy in this region. Absent from large stretches of the Downs and parts of the Weald, hares are more widespread on the Levels and marshy areas generally. Hares are also seen in woodland where they may do considerable damage by barking young trees. They visit nurseries and here, too, may make themselves a serious nuisance.

When scanning open ground with the aid of binoculars the naturalist will often detect hares lying up or feeding in areas where he may not have expected to see them. The best time for such surveys is during the early part of the year, before crops have grown up too much. Systematic observations on the hare's feeding, which can best be observed at dawn and dusk, could produce valuable additions to our knowledge of this animal.

Quite early in the year one should watch for 'mad' behaviour, 'boxing', leaping and chasing, the gathering together in companies of this otherwise solitary creature, activities heralding another breeding season. Born fully haired and with their eyes open, the young are left in 'forms'. They avoid detection by many of their potential enemies by squatting, still and silent, in these hollows in ground vegetation. Their speed and endurance, swimming ability, and detailed knowledge of their home territory help them to survive once they become independent of their mother.

Rodents

Like rabbits and hares, rodents can be both ecologically and economically important. This is certainly true of squirrels, even of our native red squirrel *Sciurus vulgaris*, a species capable of causing serious damage in conifer plantations but which many people refuse to see as anything but the innocent victim of the introduced grey squirrel *Sciurus carolinensis*.

The *Victoria County Histories* of our three counties (1902–8) record that the red squirrel was reported to be very common in some of the wooded districts of Kent, still quite common in country districts of Surrey and occasionally found within its suburban areas, and abundant throughout Sussex. As yet, the grey squirrel is not even mentioned. According to Monica Shorten (*Squirrels*, 1954), red squirrels were found in 20 per cent of Kent parishes in 1944–5, while the comparable figures for Surrey and Sussex were 11 per cent and 4 per cent respectively. The red squirrel is absent from the greater part of the South East now, the Mammal Society's provisional distribution map (April 1971) showing that since 1959 it has been found only in the Folkestone–Dover area.

Nowadays it is the larger grey squirrel that attracts attention as a frequent road casualty and because of the serious damage it does to beech and other trees. (A *Daily Telegraph* news item in 1970, 'Death leap squirrels cut power', told how they caused forty power-cuts in the South East in a year by leaping on to the wires and short-circuiting the lines!)

Of the grey squirrels introduced into Britain between 1876 and 1929, the unknown number released at Benenden, Kent (between 1892 and 1902), one hundred set free at Richmond, Surrey, in 1902, and those turned down at Frimley, Surrey, and

Sandling, Kent, in 1910, are known to have multiplied, helping to make the South East one of the main strongholds of the species by 1930.

In many places, grey squirrels are now controlled and their nests, or dreys, destroyed. Some take to holes in trees, while parks and large gardens in urban areas afford sanctuary to others. One even finds people deliberately encouraging grey squirrels in their gardens and occasionally attempting to make pets of them. In Surrey I heard of a grey squirrel following a woman into the house and begging for nuts like a dog, if ignored it seemed to attract attention by grunting several times. An uninvited grey squirrel, this time in Sussex, dashed into a house and stayed long enough to sample some fruit on the sideboard and to be photographed.

There have been other cases in the region of grey squirrels doing damage in and around houses. One Sussex housewife found grey squirrels tearing her curtains, removing large pieces of them. It is possible that this material was incorporated in the drey, which typically is made of leafy twigs and lined with grass, leaves, newspaper and other materials, in a large fork some 30–40ft up a tree.

What made another pair of south-eastern grey squirrels tear pieces off a plastic dustbin lid and nibble them is a matter for argument. These animals are known to gnaw bones, eat earth and eggs and to stalk and catch house sparrows. But acorns are the most important item of a long list of plant foods, including nuts, seeds, roots, bulbs, buds, fruit and bark. One observer claims to have witnessed an example of teamwork involving two grey squirrels, one gathering acorns among the branches of an oak and dropping them to the other which took them away and then returned for more.

Grey squirrels, like so many other common creatures, are well worth observing and there is still much to discover about them. It would be interesting, for example, to know to what extent they do actually sunbathe. Certainly groups of them have been photographed clinging to trees while apparently engaged in this activity (or should one say 'inactivity'), some facing towards the ground and others in the opposite direction.

Like the grey squirrel, the dormouse *Muscardinus avellanarius* is arboreal in its habits. Once a common pet, this delightful little mammal has declined in numbers, but south-eastern naturalists should have no real difficulty in finding it in hazel coppice, hazel hedges, or oakwoods where there is plenty of scrub. Here stripped stems of honeysuckle will show where dormice have gathered bark to make nests, which may be near the ground among hazel shoots or stems of saplings. Nesting boxes intended for birds should also be carefully examined, for dormice have been found in them as much as 12ft from the ground.

Mouse-size, bushy-tailed, bright yellow-brown above and white below, the dormouse is both silent and nocturnal. Prior to hibernation, or winter-sleep, which has lasted as long as six months and twenty-three days, it fattens itself on hazel nuts, beechmast and other tree seeds. Any harm that dormice may do here is generally considered to be almost negligible. However, in Czechoslovakia, where they were stated to be very common after good mast years when trees produced abundant fruit and seeds, they were reported to do serious damage to young trees by nibbling the shoots and bark.

Little is known of the dormouse's natural predators, but cats have been seen to catch them in our region. A weasel was thought to be responsible for the destruction of dormice in one Sussex locality, while many of those that were a noteworthy feature of the fauna of Ashdown Forest and elsewhere un-doubtedly perished in heath and woodland fires.

The wood mouse (or long-tailed field mouse, as it is also known) *Apodemus sylvaticus* is common and widely distributed in the South East, being found not only in woodland, gardens and fields, but also in houses and garden sheds in winter. The wood mouse sometimes outnumbers the house mouse *Mus musculus* in buildings in winter, though this latter animal is still often common in and around farm buildings.

The wood mouse spends much of its short life in runways in

and below dead leaves, moss and other vegetation covering the floor of woods and other undisturbed ground. It makes a nest of shredded grasses below ground and rests and breeds there. Usually emerging at night to forage, it is sometimes seen in daylight, perhaps as it collects nesting material in spring or summer, or as it strips bark from woodland shrubs in wintry weather. It stores food below ground in winter, but at other times it makes great use of old birds' nests as feeding platforms. Along one stretch of hedge, some 100yd in length, there were six birds' nests filled with chewed hips and haws. One of the nests, a song thrush's, had been domed and filled with chewed berries, the entrance hole being at the bottom.

In addition the wood mouse will take certain farm and garden plants. Sprouting winter wheat may be grazed, sometimes seriously. Newly planted peas are uncovered and eaten. Crocus corms and tulip shoots are devoured, while buds and flowers of the Christmas rose have also formed part of this animal's diet. One enterprising individual, which made a winter nest of newspaper, polythene and baling twine in a caravan, ate a quarter of a bar of soap; another fed on wallpaper. It is perhaps just as well that this prolific little rodent is preyed upon by cats, barn and tawny owls, and small carnivores.

Some writers consider the wood mouse and the yellow-necked mouse *Apodemus flavicollis* to belong to one and the same species, but this suggestion is usually dismissed on the grounds that forms intermediate between them are rarely found. A larger and brighter coloured animal than its close relative, the yellow-necked mouse's chest-spot spreads to form a broad collar. In our region it lives in the same kinds of places as the wood mouse.

Breeding has taken place in compost heaps in Sussex, a nest with eight young having been turned out of one in July and another with seven in October. In the same county the species has been found nesting in apple stores, and in one such place a yellow-necked mouse tore up paper and sacks and imported half a bucketful of acorns before being caught. I used to trap yellow-necked mice in my house, but never found their nests there.

Much more active during the daytime than other mice, the harvest mouse *Micromys minutus*, the smallest British rodent, survives in the South East. Though a Sussex field of only 3½ acres held fifty nests as recently as 1965, the harvest mouse's ball-shaped nest of woven grasses is nowadays best sought in field hedgerows, reed beds and woodland margins. At Pagham nests are sometimes found in the reed beds below the sea wall, while elsewhere they are seen in temporary reed stacks, in sedges on river banks, and in coarse vegetation. One nested in a dahlia bed in a garden, and another built on a warbler's nest 15in off the ground.

Though the harvest mouse is generally reckoned to live 16–18 months in the wild, one was kept in captivity for 32 months. Although much remains to be discovered about the harvest mouse, especially in connection with breeding and feeding, we do know that it is taken by cats and barn owls and that any harm it does appears to be almost negligible. Certainly anyone who has observed the ease and speed with which the harvest mouse runs up the stalk of a wheat plant, and the use it makes of its long prehensile tail, will hope that there will always be a place for this tiny, dainty creature in the countryside.

One cannot express these feelings for rats, extremely destructive carriers of diseases affecting man and certain domestic animals. The black rat *Rattus rattus* is largely confined to well-established colonies in ports on the fringes of the region. The brown rat *Rattus norvegicus*, despite numerous attacks made upon it, persists in hedgerows, rubbish tips and ditches, and there are some large colonies along the coast, where food is found among tideline debris. Reports of 'plagues' usually refer to rats moving from cornfields in autumn or to those seeking shelter in outbuildings.

Apart from man, brown rats have several natural enemies. Barn, tawny and short-eared owls will devour them, as analysis of the remains ejected in their pellets makes evident. Kestrels will also take them, as will foxes and certain cats and dogs. But these predators do not effectively control the rat. People destroy vast numbers of rats, but sometimes the very same folk encourage the increase of those remaining. Thus destruction campaigns launched against wood pigeons in winter have resulted in numerous carcases of this voracious pest being dumped unburied in hedgerows where rats have dug themselves in, moving on only when the food supply was exhausted. Food put out carelessly for birds may encourage rats, as may inadequate storage facilities in both town and country.

When feeding on the remains of creatures killed on the roads, where they, too, are frequent victims, rats may be aptly described as scavengers. The same is true when rats live beneath large starling roosts, burrowing in under inches of droppings and feeding on dead and dying birds that are dragged into the entrance of the tunnels. At spawning time, brown rats visit ponds and kill and eat large numbers of frogs and toads.

As well as destroying, damaging and contaminating crops and stored foods, rats will devour insects, eggs, young birds, and other small animals and their young. They have even been known to remove and eat putty from newly erected window-frames, probably being attracted by its linseed oil content. This omnivorousness has contributed greatly to the brown rat's success, as have its fecundity, hardiness and adaptability. Though the wild brown rat is a carrier of the disease organisms of food poisoning (Salmonella), leptospiral jaundice and trichinosis, its domesticated strains are of great importance in medical research. When I watched a great tit pulling hairs from a dead rat for nest building, I was reminded that even when it is no longer living an organism may play a part in the complex balance of life.

The field (or short-tailed) vole *Microtus agrestis* is another widespread member of our regional fauna whose activities are not without significance. Though often ignored, this stumpy-tailed, mouse-size creature does attract attention when its numbers reach one of their periodical high peaks and the word 'plague' may sometimes then be used with some justification.

Living in a network of runways at, or just below, ground level in rough grassland, meadows and marshes, gardens and open places in woods, the field vole may suffer heavy losses of young in wet weather. The adults are sometimes able to escape floods by making use of their ability as swimmers. One was seen to swim at least ten feet from one grass tussock to another on a flooded marsh, while another was observed swimming across a pond in the comparative safety of a garden. Snow is less of a hazard to voles. Indeed it forms a blanket between them and certain of their enemies, and dead grass beneath this cover often becomes honeycombed with vole runs.

The field vole feeds mainly on the bases of grass stems and the presence of bright green droppings and heaps of cut grass stems indicates that a runway is currently in use. In Sussex the field vole has been blamed for causing damage to lawns and among young lettuce plants. This animal has also been held responsible for ring-barking and killing over half of the young trees in a pine plantation.

Becoming sexually mature at three weeks, the female mates at six weeks. Three weeks after mating the young are born in a nest of grass under a grass tussock or below ground. The breeding season is generally considered to extend from March to September, but I have heard of a case where reed-cutters found two newly born young as late as 10 December. They were dead and were believed to have been killed by the frost of the preceding night. Apart from men and owls, the enemies of the field vole include cats, dogs (including the dachshund) and foxes, whose method of pouncing on them is fascinating to watch.

In the South East remains of the bank vole *Clethrionomys glareolus* seem to be less commonly found in owl pellets than those of the field vole. An agile climber, the bank vole is at home in woods, banks, hedges and Downland scrub. It eats insects, other bank voles and various plant foods. The vegetable foods include fungi, shallots, bulbs and crocus corms, and the species was believed to have been responsible for the destruction of parsnips and fodder beet which were eaten out and reduced to hollow shells about an eighth of an inch thick.

Rat-size and misleadingly referred to as the water rat, the water vole *Arvicola terrestris* may still be seen by quiet and patient observers as, hunched-up, it feeds on grass by ponds and streams. Also an inhabitant of rivers, mill ponds, coastal marshes and brackish dykes, the water vole has suffered from the disturbances of drainage operations and the destruction of waterside vegetation. A target for ignorant people out shooting, it is also taken by owls, pike, American mink and herons. That some escape these enemies is due mainly to their ability to remain motionless among waterside vegetation, to dive with a gentle plop to safety and to float quietly in the water until danger passes.

In recent years careful observation of the water vole has produced a number of accounts of its feeding and other habits. Particular interest attaches to reports of water voles climbing trees to feed, in one such case the animal having collected pond snails and eaten them on a branch of an alder overhanging the water, while in another a water vole had climbed the upper branches of a hawthorn about 12ft from the ground where it was watched devouring newly burst buds.

Water voles have been seen barking the roots of apple and pear trees in March, and they have been recorded as killing young poplar trees by eating the underground roots. These animals have also been known to strip bark from full grown willow trees and to eat the leaves of field maple. Iris and reedmace are among other plants taken by the water vole whose habit of 'mowing' a strip of waterside grass yields a regular supply of young growth. In a garden, where there was a temporary 'plague' of these animals, water voles burrowed under sweet william plants and tulip bulbs, storing many of the bulbs in their burrows.

The burrows are often used for breeding, the young being born in nests made of rushes and grass. Sometimes, though, similar nests are constructed in reed-beds and other places above ground. Burrows and nests form part of the animal's home

Page 53 South Downs, Sussex; (*above*) looking towards Chanctonbury Ring (783ft) from the Devil's Dyke near Brighton; (*below*) Mount Harry and Firle Beacon from Black Cap; scrub is present and chalk is exposed in places

Page 54 Outcrop near West Hoathly, Sussex, of the massive sandstone which forms the upper part of the Lower Tunbridge Wells Sand in the western High Weald

range, an area that is marked by a secretion of the flank musk-glands which is smeared on to the hind feet and then stamped into the ground. As part of this home range they are defended by the male whose aggressive posture, chattering and fierce fighting spirit help to drive away intruders. In Britain the water vole's burrows do not cause much real damage to river banks.

The musk rat *Ondatra zibethica*, an imported species whose fur is the musquash of commerce, is capable of doing great damage to river banks. Fortunately its colonies in Surrey and Sussex, like the larger ones in Shropshire and Scotland, had been exterminated by 1937.

Another introduced species, the coypu or nutria *Myocastor coypus*, was the subject of a thorough control campaign before it had the chance to become a really serious threat to river banks by burrowing. Though never as numerous here as it was in the Norfolk Broads district, this South American rodent was sighted in a number of places in the South East during the period 1960–70. Large, somewhat rat-like, with prominent orange-yellow incisors, the coypu is thoroughly at home in reed-beds, marshes and along river banks. Its presence should be reported to the authorities immediately because isolated colonies quickly increase in size and the resultant damage to pasture and crops can be considerable.

Cetaceans

Following the order of the *Handbook of British Mammals*, we arrive at the Cetacea, the group including whales, porpoises and dolphins. Here, in a region whose long stretch of coast is accessible at so many places, are opportunities for naturalists to make useful contributions to our scanty local knowledge of mammals so well adapted to an entirely aquatic existence.

The common porpoise *Phocaena phocoena*, the commonest and smallest British species of cetacean, may be seen offshore and occasionally individuals come into harbours like those at Pagham and Chichester. Unlike the porpoise, the common dolphin

D

Delphinus delphis, another fish-eater, is speedy and fond of leaping clear of the water. 'Schools' are observed off the coast, while examples are sometimes stranded on the shore. Movements of Risso's dolphin *Grampus griseus* off the Sussex coast have also been reported in recent years.

Carnivores

Both the common seal *Phoca vitulina* and the grey seal *Halichoerus grypus* are seen on the South-East coast, though they are not known to breed here. In Kent, the common seal occurs in small numbers both at Sandwich Bay and on the Swale, while in Sussex individuals have been observed offshore at Selsey Bill, in Pagham Harbour and off East Head (Chichester Harbour). Early in January 1968, an immature female common seal 3ft 5in long was seen well inland, in the river Arun, at Greatham Bridge; sad to say, it was shot.

In recent years, grey seals have been sighted in Chichester and Pagham Harbours, at Cuckmere Haven, and offshore at Fairlight. A young one was stranded on the beach at Goring where it was later joined by two adults. There are also non-breeding records from the Kent coast.

Naturalists should not attempt to rely on colour or size in order to distinguish these species. They should look carefully at the silhouette of the seal's nose, which is concave in the common seal, but convex (male) or straight (female) in the grey seal. Another important feature is the shape of the nostrils, those of the common seal forming a 'V' shape and almost touching below, those of the grey seal being almost parallel and not touching below.

Feeding on fish and shell-fish, seals have long been the subject of complaints, controversy and persecution in places where they are more numerous than here. In the South East other carnivores have attracted such attention.

Early this century a contributor to the *Victoria County History of Surrey* stated that little mercy was shown to the badger *Meles*

meles if a chance was obtained of shooting or catching it or even running into it with a pack of hounds. Nowadays there are reports from time to time of attempts to destroy badgers by gassing, an illegal practice. 'Brock' also suffers as a road and rail casualty, while six (and very probably another six) of seventeen badgers, whose deaths in south-eastern England were investigated by the Nature Conservancy, died of dieldrin poisoning.

Despite such losses, though, this largely nocturnal animal is thriving in many parts of our region. Badgers are even seen in busy built-up areas where some have become so tame that they make regular visits to houses and gardens for food. At Mile Oak near Brighton, people were accustomed to the sight of badgers, one of which had been seen walking along a garden wall. But it was not until workmen found two badgers in a set beneath the foundations of a wooden house, which was being demolished, that they realised how established as local residents these animals were.

In those areas where people tend to regard neatness and tidiness as being, like cleanliness, next to godliness and virtues in themselves, the badger is sometimes given a bad name (and poisoned food) because of its habit of searching for grubs and earthworms, a most important item of its diet, in lawns, herbaceous borders, sports grounds and golf courses. Some people have been able to discourage badgers from grubbing little holes in their lawns by scattering a few scraps there in the evening, bits of pastry being highly recommended for this purpose.

Foresters, thankful for their destruction of small rodents in the woods, often instal special badger-gates, enabling badgers to come and go without forcing up wire-netting fences. Poultry farmers and gamekeepers sometimes complain about badgers, but one Sussex poultry farmer claimed to have shared his land with a large colony of badgers for thirty years without ever losing any stock and there are woods where pheasants and badgers happily co-exist. Unhappily the occasional senile 'rogue' badger with teeth missing or badly worn-down takes to raiding poultry or game. R. J. Jennings, who a few years ago sent photographs of skulls of badgers in good health and in old age to *The Countryman*, stated that in twenty-five years he had reluctantly destroyed four adult badgers after being presented with indisputable evidence that they had taken to raiding poultry runs and hen houses from woods of which he had charge. All four were senile, their remaining teeth being so broken and worn that they could not have obtained food normally.

Few people would object strongly to the humane destruction of badgers in such comparatively rare cases. What they do object to, and rightly, is the all too frequent disregard of the fact that generally badgers do more good than harm, and are of little economic importance, in areas where their natural foods are plentiful. In addition to the grubs, earthworms and young mammals already mentioned, these foods include apples, acorns, grass, fungi, snails and insects.

The complexity of the whole question of an animal's food can be illustrated by considering the wasp whose grubs are eaten by badgers, which avoid many stings by digging, usually by night, from above the wasps' nest. People living near wasps' nests often regard the insects as pests and one fruit-growing and jam-making firm I know will destroy nests in the district without charge because of the damage wasps can do to ripe fruit and also because their presence is unwelcome in the factory. On the other hand, wasps feed their larvae on other insects and thus destroy many pests during the course of a summer.

The relations of badgers with foxes, important predators of poultry and game in some areas, are not simple either. For while it is not uncommon to find 'Brock' sharing his set with foxes (not to mention rabbits and rats), he will occasionally kill fox cubs and, much more rarely, adult foxes. The trouble may sometimes arise through pressure on living space within the set, but this burrow system is usually roomy, consisting of many tunnels, chambers and entrances. Dug in a natural bank, valley side, wood or pit hole, the badger's complex labyrinth of tunnels and chambers is often marked by enormous mounds testifying to its industry and excavating ability.

The set is kept clean and tidy, and naturalists who cultivate the art of watching these largely nocturnal animals with the aid of a red torch may be rewarded by the sight of badgers spring-cleaning, casting out soiled bedding and stale earth, or bringing in fresh grass or bracken. A most satisfying experience for the observer is when he is able to enjoy the sight of lively badger cubs at play outside the set and to listen to them leaping through the grass. But perhaps the most exciting experience is when, provided with raisins and rusked bread (stale bread baked to crispness in the oven), the naturalist wins the confidence of the cubs, and has them feeding from his hand and wagging their tails like friendly, contented dogs.

Born during the first five months of the year, often in February in the South East, badger cubs remain in the set for some two months. When they are about three months old they take semi-digested food regurgitated by the mother. They soon start to search for food independent of the sow, but live with her until the autumn and sometimes until winter has passed. Their subsequent movements, like those of adult badgers, follow the urges of the mating season, the state of the food supply and the threat of danger. Of great assistance in these travels is their ability to set scent trails by means of oily musk-scented liquid secreted by glands in the anal region.

Unlike the badger, the fox *Vulpes vulpes* is an untidy creature whose earth is often littered with feathers, bones and other remains of its prey. In the countryside hunting, cub digging, gassing, shooting and snaring all go on, but Reynard is still a common and widespread inhabitant of woods and scrub. In built-up areas, where many of them are finding food and safety, foxes are seen occasionally in daylight and more commonly in headlights at night.

Much of the controversy concerning foxes has involved their prey which, as I have already mentioned, included many rabbits and still does include such other small mammals as rats and voles.

The trouble arises when foxes leave such wild pests and take to attacking litters of young piglets, as one did in broad daylight at one place in Sussex, or to killing reared mallard, as foxes did elsewhere in the same county. In the preserved areas of western Kent pheasants and partridges were important items in the diet of foxes. Yet so complex are the inter-relations of different species that one hears of a fox rooting for grubs being seen to be followed by 14 pheasants in December.

This liking for grubs leads foxes, sometimes in family groups, to dig in middens and to nose under cowpats and other animal droppings, while their taste for earthworms results in garden compost heaps being disturbed and in nightly visits to lawns where succulent worms are lapped up before they can retreat into their burrows underground. In late summer and autumn Reynard may eat toadstools and, early in the morning at this time of year, the fox may sometimes be seen blackberrying as it stands on its hind legs against the bushes or clings half-way up a hedge.

Ever ready to vary its diet according to place and season, the fox shows an adaptability which, together with its ability as swimmer, runner, climber and digger, often ensures its survival when every hand seems to be turned against it. During hard winters, when numerous other mammals may perish, foxes may be helped by their readiness to feed on carrion. Certainly their will to live is illustrated by the case of the foxes living in a sand-pit which enlarged the nest-holes of sand martins and dragged out the contents while actually clinging to the vertical face of the sand.

Because of the animal's cunning and determination many countrymen feel a sneaking regard for the fox, but it is vital to remember the destruction it can cause. Equally one cannot over-look the fact that, though most species of mammals are potential carriers of the disease, the fox is particularly susceptible to rabies, which is fatal in man.

Apart from two fairly recent cases of rabies in imported dogs in 1969 and 1970, no animal rabies has occurred outside quarantine kennels in Great Britain since 1922. The 1969 case of rabies involved a dog out of quarantine at Camberley, Surrey, where the Ministry of Agriculture & Fisheries decided to try and eliminate the larger mammals and carrion-feeding birds which might have come into contact with it. In the result a comparatively small number of potential carriers of the disease was destroyed. If no action had been taken and the disease had been transmitted to wildlife many more animals would have had to be slaughtered over a much larger area.

While admiring its decisive action in the Camberley case, some naturalists feel that the Ministry is not doing enough to control feral American mink *Mustela vison*, potentially destructive animals found near water in several places in the region. Descendants of individuals that escaped from fur farms started in Kent and Sussex from about 1929, these mink are naturally dark brown with a small white spot on the lower lip and chin, but light-coloured individuals are occasionally seen. Expert swimmers and climbers, they breed successfully in the wild here, living on water birds and water voles. Five established themselves in hollow willows at Woods Mill, Sussex, and sixteen were caught in the lake at Knepp Castle, and there are fears that the species may increase in places where trapping is not carried out.

It is important not to confuse these mink with the otter *Lutra lutra* whose scarcity in the South East is giving cause for concern. It is important, too, that visitors to rivers and marshes who observe the otter or hear its contact call—a short whistle which may be repeated at intervals—should think carefully before imparting the information to anyone but a responsible officer of a county naturalists' trust.

An expert fisherman, the otter seems to be especially fond of eels in some areas. Its taste for this species may, it has been suggested, be useful in trout and salmon streams where eels devour eggs of trout and salmon. The otter's diet also includes frogs and tadpoles, rodents, birds, beetles, slugs and earthworms. But game fish, salmon and trout, may also form part of this mixed diet and this, together with the otter's habit of 'sampling' a fish, taking a mouthful or two from the back near the head and leaving the rest to rot, has involved it in controversy from time to time.

Otters have long been hunted, but there is no evidence that hunting has had any controlling effect on the species in the South East. But, at the same time, one heartily agreed with the view of *The Sussex Mammal Report* (1969) that, 'While numbers are so low it would have been more realistic to abandon all hunting in Sussex and give the otter a chance, if it has one, to recover.' This viewpoint did not receive complete support, but one was encouraged to read in the same publication that the Arun and the Medway would not be hunted during the next few years and that other rivers would be hunted but no kill made unless considered necessary.

In a symposium on Britain's beasts of prey, organised by the Fauna Preservation Society and the Council for Nature, the decline in status of the otter was attributed to the dredging of rivers, the effect of toxic chemicals and river pollution, and regular trapping for skins. One cannot be certain that all of these causes apply in this region, but we do know that many habitats have been destroyed as the result of drainage schemes. Otters are recorded as road casualties and I have heard of several being killed at one place since the culverting of a stream which formerly went under a bridge.

While we obviously cannot guarantee complete protection to the otter or indeed any other wild species, we can, by supporting the nature conservation trusts, help to provide more reserves where animals and plants of all kinds are welcomed by man the benefactor and are protected from many of the activities of man in the capacity of destroyer. And it is pleasing to record that the Sussex Trust's headquarters at Woods Mill has provided sanctuary for otters on a number of occasions.

Certainly those who have watched otters dread the very thought of the species vanishing completely from our regional fauna. For there can be few wild mammals as capable as the otter of making so plainly obvious what can only be described as a sheer joy of life and living. When being taught by their mother how to dive and swim, how to fish and avoid danger, the two or three cubs make a lively group. Playing in the water and tossing and catching sticks and stones are all part of the fun. Adult otters can be just as playful and slides where they toboggan down steep banks are sometimes seen by rivers. In winter they will roll in the snow, make slides in snow or slush or on the ice and playfully chase away coots and other birds that may venture on to their playground.

Their general build and appearance mark the stoat and the weasel as members of the same family as the otter. But what often surprises people is that these smaller carnivores are as capable of play as the otter and other larger members of the order Carnivora. The antics of stoats have to be seen to be believed, particularly when they are playing in groups, chasing each other, leaping into the air, turning somersaults and boxing and wrestling together. Weasels also have their less serious moments and one may be lucky enough to observe a family party playing in a bundle and jumping on one another.

The stoat *Mustela erminea* and the weasel *Mustela nivalis* are relatively ancient members of the British fauna. Both occur in the South East on farmland, in woodland, in water-meadows at the coast and along sea-walls, while their corpses may often be seen on game-keepers' gibbets. For, in addition to the foods mentioned earlier in this chapter, stoats will take game birds and their eggs and chicks, while weasels, although they generally prefer small mammals such as rats, mice and voles, have been known to kill pheasants in the region.

In recent years there have been several reports from the South East of stoats 'in ermine', the condition that follows the autumn moult, when the animal becomes white except for the black tip of the tail. Even when the winter was not severe individuals in partial ermine were seen. This condition also occurs in the weasel in the north of its range, but rarely in Britain, though there is an old winter record of a Kent specimen whose coat had turned white along the back.

Deer

Recorded in Domesday Book as being well established in many parks, the fallow deer *Dama dama* is well represented in the woodlands of our region. Several feral herds were established when animals escaped from deer parks and it has been found necessary to control the numbers of this deer in certain areas.

Standing about 36in high at the shoulder, fallow deer have a summer coat which is typically deep fawn on the upper parts with numerous white spots. Additionally, though, this species includes many colour types. In Sussex there are several scattered groups of a dark variety, while white ones have been reported from St Leonard's Forest and other places in the same county.

The buck's 'groaning' call is heard in October, when the sexes come together for the mating season or rut. Then the bucks gather the does into harems in traditional territories known as rutting stands, driving off rival males, activity that may lead to fights and broken antlers. After the rut, which lasts about four weeks and continues into November, the separate buck and doe herds start to re-form. The fawns are born in May or June but remain in cover until they are about three weeks old, when they start following their mothers. It is then that the silent observer is rewarded by the sight of fawns playing and gaining strength.

Fallow deer feed mainly on grasses and small plants, but when they start bark-stripping, a practice that is sometimes believed to be confined to females, they can be troublesome, damaging beech and other trees up to a height of six feet. In winter they may be seen feeding on moss uncovered by scraping through the snow.

Until recently roe deer *Capreolus capreolus* in Britain were not known to resort to the gnawing or stripping of bark, but this habit has now been observed on lodgepole pine in several different places. In suburban districts of Surrey and Sussex the roe deer's fondness for many kinds of garden plants, including privet and roses, has been the subject of complaints. Persistent browsing by these deer on young Norway spruce results in shapes like those produced in topiary work. This effect of the animals' nibbling of shoots and buds has also been observed on yew bushes, some roe deer apparently developing an immunity to yew poisoning, a state which can lead to the death of deer and other animals. When roe deer graze grass and small plants in woods they may prevent the natural regeneration of forest crops by destroying small seedling trees. On the other hand, such grazing may help to reduce fire hazard in woods and rides by keeping down vegetation. Normally dainty feeders, roe deer may be seen ploughing deep furrows in snow by running with their shoulders close to the ground and then using their forefeet to scrape down to vegetation that has remained unfrosted.

Unlike the red deer and the fallow deer, the smaller and graceful roe deer, a woodland species, is generally monogamous, the bucks making no attempts to gather harems. Though roe themselves are often difficult to find, the presence of frayed trees, whose bark has been rubbed off, will show where the buck has marked his territory, an area that is usually well established by the second half of April. A worn or beaten track round a tree, a tall tuft of vegetation, or some similar focal object, may well be a rutting ring where the buck chased the doe during the rut in late July or early August.

Roe deer certainly seem to be spreading and increasing in the South East. But the red deer *Cervus elaphus*, our largest British species, now gives regional naturalists much cause for concern. At Ashdown Forest which, despite its name, is mostly heathland, a once fine herd has virtually disappeared. But perhaps there is some consolation in the fact that the finest herd of park red deer in the country is at Warnham Court, Sussex, whose park has long been renowned for its remarkable stags.

Muntjac *Muntiacus reevesi* which in their wild state seem to have originated from animals that escaped from collections at Woburn and Whipsnade are found in West Sussex and possibly in other parts of the South East, though their small size—they stand 17–19in high at the shoulder—and fondness for dense undergrowth makes them difficult to detect. The muntjac differ from the other deer mentioned in that they appear to breed throughout the year. Recorded from the north of our region, the introduced sika deer *Cervus nippon* is another species that may often elude observers. Fond of dense forestry plantations and woods with plenty of undergrowth, sika are similar to red deer but smaller in size, standing 32–35in high at the shoulder.

Regarded by some people as a means of adding to the wealth and variety of our fauna, the introduction of such species as the muntjac and the sika deer, whether deliberate or accidental, is frowned upon by others. The absence of such larger predators of deer as wolves sometimes leads to fears that a species may increase its numbers beyond bounds and become a serious pest.

Some deer do die of starvation in severe winters and others may be so weakened by starvation that they succumb to the attacks of parasites. Foxes take some young and uncontrolled dogs are responsible for other losses. But what the naturalist may have to bring himself to accept is systematic and humane control of deer by scientists who have studied the animals' habits and are assisted by trained and skilled marksmen. Absence of such proper control has, all too often in the past, resulted in wounded deer being left to limp away and die and in others surviving but with pellets embedded under the thinner areas of the skin.

The word 'control' is, of course, alone sufficient to incur the wrath of many folk. But every case of deer wounding, badger gassing, or of bats being starved to death, simply emphasises that much more thought must be given to the whole question of the status of wild animals. In such a highly developed area as the South East forms of management, involving degrees of control of numbers, may well prove more acceptable and effective than the extremes of unbridled persecution and feckless neglect of our wild mammals.

CHAPTER THREE

The birds

*Birds of prey—Game-birds—Water-fowl—Waders—
Sea-birds—Pigeons—Crows—Other birds—Ringing
and migration*

'I ONLY SAW a few larks and whinchats, some rooks, and several kites and buzzards.' Thus wrote Gilbert White of Selborne from Ringmer, near Lewes, in 1773. He was visiting the Sussex Downs, his 'chain of majestic mountains'.

*Birds of prey**

Now, close on two centuries later, the kite, once resident in the South East, is a very rare and irregular visitor. The buzzard is far from common, but it can still be seen soaring and drifting over the Downs and in other parts of the region. Taken by collector and gamekeeper alike, it had a sad history of persecution before the 1939–45 war, and as recently as 1968 one bird of a pair was shot in Sussex, albeit accidentally.

In recent years buzzards have been observed breeding, or attempting to do so, in Surrey, where nine were released in 1939 in Witley Park, near Godalming. Two of the five pairs present in Sussex during the 1967 season are known to have raised young but offspring do not appear to have been seen in subsequent years, though breeding may well have taken place.

The hobby is another bird of prey whose breeding numbers are being watched with keen interest and not without anxiety. This agile falcon, a fine interceptor of larks, pipits, swallows,

* Scientific names of birds mentioned in the chapter will be found in the *Index of South-Eastern Birds*, pages 182–98.

67

martins and certain larger birds, has bred in each of the three counties in recent years. In Surrey, pairs have nested in an old crow's nest and in a nest built on a squirrel's drey.

Breeding in the wilder places, hobbies may sometimes escape observation by bird-watchers, and anyone who encounters a breeding pair of this or any other threatened species should inform his local bird club. Certainly every effort should be made to alert responsible local people to the need for vigilance. Thus it may be possible to prevent such sad destruction as happened in 1960 when a Kentish gamekeeper shot an osprey, thinking it was some other species. The same year exhaustive measures had been taken by the Royal Society for the Protection of Birds to allow ospreys to breed in Scotland.

That such watchfulness is necessary is evident when one considers how badly the kestrel suffers near built-up areas from the attentions of nest robbers, who take both eggs and young. Despite this persecution, though, the wind-hover, as this long-tailed falcon is also called, is well on the way to re-establishing itself securely in the South East.

The kestrel's nesting habits show that it is an adaptable species. In some areas it nests in ancient hollow oaks, as in Richmond Park, or makes use of old crows' nests in tall trees standing in commanding positions at the edge of woods. It breeds successfully in Brighton and other towns, using ledges on high buildings, chimneys, steeples and towers. Incidentally, church councils could encourage the spread of this bird (and also the barn owl) by removing wiring from towers and using other methods to discourage jackdaws and pigeons; if this is thought to be really necessary.

The kestrel is equally at home in some of the Downland chalk quarries and still nests along the chalk cliffs, though there has been a marked decline in this latter situation. At Dungeness a few pairs breed on pylons near the nuclear power station.

The kestrel's choice of food is by no means restricted to small rodents. Young house sparrows and other small birds are often used to feed the nestlings, while bats, frogs, insects, snails and worms have all formed part of the kestrel's diet. To secure their food, kestrels search open land such as commons, parkland and grassland. Corn stubbles are favoured after harvest, but such areas as farmyards and road verges are not neglected.

Much the same is true of the barn owl, though it does seem to prefer farms where buildings provide nesting sites and open fields make excellent hunting grounds. Though this bird is normally nocturnal, it is occasionally possible during daylight hours to watch it 'beat the fields over like a setting-dog' (who could resist Gilbert White's picturesque phrase!). Slowly and methodically, silently and close to the ground, the barn owl searches for such small mammals as voles, shrews and long-tailed field-mice. At other times small birds, beetles, moths and even frogs may be taken.

One can establish that these creatures are included in the barn owl's diet by examining the compact, blackish pellets of bones, fur, feathers and other indigestible material, which the bird coughs up at its regular perches. Barn owl pellets from Richmond Park and Bookham Common showed a preponderance of field voles in the bird's diet. At Sedlescombe, barn owls caught one bank vole to fifty-two field voles. At Amberley Brooks barn owl pellets contained a few common shrew skulls.

That this useful bird, with its characteristic white heart-shaped face and beautifully marked light-buff plumage, should have been ruthlessly persecuted is impossible to believe. Unfortunately, however, such senseless destruction did occur and is, no doubt, one reason why barn owls, though widespread, are no longer plentiful in the South East. The species has also suffered from disturbance and loss of habitat, while toxic chemicals used in the countryside have taken their toll. As several individuals have been found dead by the roadside, it is likely that fast modern traffic constitutes yet another hazard to barn owls.

At one time the decrease of barn owls in certain places where formerly they were common was attributed to competition from the little owl, our smallest British owl, though this was never established as a fact. On the other hand, however, there is an interesting record of a barn owl and a little owl being flushed from the same place and pellets of both being found in the same hole.

The little owl is now widely distributed throughout our region as a resident breeding bird, a state of affairs stemming from the release of forty imported little owls by E. G. B. Meade-Waldo at Stonewall Park, Chiddingstone, Kent, between 1874 and 1880, the turning down of two further batches, each of twenty-five, at Edenbridge in 1896 and 1900, the introduction of more birds at East Grinstead, Sussex, in 1900 and 1901, and possibly some additional but unrecorded releases.

One pair of these introduced little owls bred near Edenbridge in 1879 and by the end of the century Meade-Waldo was able to record regular breeding. With the bird's continued spread, a storm of protest arose from people unaware that Charles Waterton had originally imported several little owls from Italy because gardeners there valued the species 'for its uncommon ability in destroying insects, snails, slugs, reptiles and mice'.

In 1935 the British Trust for Ornithology agreed to promote a thorough inquiry into the little owl's food and economic status. From the *Report of the Little Owl Food Enquiry 1936–37* a number of facts emerged. Insects and rodents taken on the ground, where the little owl is adept at running fast, formed the main food at all seasons. There was no evidence of prey being killed in excess of immediate requirements or of carrion being used in any significant quantity. During the breeding season, rats, rabbits, voles and mice were taken, as were starlings, house sparrows, blackbirds and thrushes. Other birds (including game chicks) were rarely taken in normal circumstances, except by the occasional 'rogue' little owl, and, while those in holes were raided occasionally, nests were generally left alone. Naturalists could help to keep our knowledge of the little owl's feeding habits up-to-date by examining pellets and submitting lists of the contents to local natural history societies.

Now protected by law, this noisy species may often be seen

Page 71 Pine and birch in Brantridge Forest, Sussex; these trees are active invaders of heathland in the South East

Page 72 The South Downs, Sussex, above Alciston. In the distance is Windover Hill, which is probably the site of an ancient flint-mine

flying by day, though it hunts mostly at dawn and dusk. Its nesting sites include holes in trees, walls and the ground, while one pair has nested on the roof girders at a gas works. Nests have also been found under girders of bridges and in the side of a haystack.

Three other types of owl still nest in the South East. The tawny owl, whose breeding success depends largely on the numbers of such small mammals as bank voles and wood mice, is the most generally distributed of these. In places it is abundant, Richmond Park having a particularly high breeding density. It prefers well-wooded districts where old trees provide secluded nesting sites, and where its deep far-sounding hoot reveals the presence of this nocturnal species.

In the early part of the century the long-eared owl was not uncommon in certain parts of the region, but there are few breeding records now. In Kent this bird, which often takes over deserted magpie nests, nested in an osier bed at the edge of a marsh. Pellets found in a stunted conifer wood in the same county contained skulls of the common shrew. The short-eared owl is another species that bird-watchers should keep a careful eye on. It has been actively persecuted in recent times, but a few pairs still nest on the Medway estuary.

Now fully protected throughout the year, the sparrowhawk not only fell to the gun in large numbers but suffered when its prey, small birds, had taken seeds dressed with poisonous chemicals. A bird of woodland and wooded farmland, it remains scarce in the South East, where the number of young reared appears to be small.

Before leaving the birds of prey, I would direct the reader to the alphabetical *Index of South-Eastern Birds*, pages 182–98, where he will find the names of other species which await the bird-watcher who is equipped with binoculars and note-book.

E

Game-birds

The harassment and destruction of birds of prey in the South East and elsewhere has often been done in the name of game preservation.

The pheasant, the attractive long-tailed species that has been so carefully safeguarded by gamekeepers, is well distributed in our area. That it is ignored by many naturalists is a pity, because it is an interesting bird. Largely a woodland species nesting under bracken and other ground-cover, the pheasant is also attracted to marshes, while a few still survive from earlier introductions on the shingle area at Lydd.

The diet of this gluttonous bird is full of surprises for anyone prepared to study crop and gizzard contents, and to reflect on the ways in which the pheasant's feeding habits affect other living things. Plant food swallowed includes acorns (seventeen in one crop), hazel nuts, wheat grains (759 in one crop), small potatoes, tubers of pignut and pilewort, beechmast, holly berries, hawthorn pips and weed seeds, to name but a few. Animal food has included insect larvae (2½oz of St Mark's fly larvae from one bird), snails, slugs, spiders, beetles, woodlice, small adders (eight baby adders in one pheasant's crop), field mice and voles.

The pheasant was introduced into Britain before the Norman Conquest. The common partridge, however, was originally native to our country, but many continental birds have been brought in and released during the present century. A successful bird of cultivated land, it appreciates the hedgerows and rough bushy patches and grassy corners that are tending to disappear in many places. As a breeding species the unobstrusive common partridge is widely distributed in the South East, where it may be seen flying fast and low over farmland, marshland and downland. Its diet includes stubble-corn, grass, clover, greens, flowers, buds, weed seeds, insects, grubs, ants and their pupae. In autumn and winter partridges occur in family coveys, but these disperse early in the year.

Unlike the common partridge, the red-legged partridge is given to running, rather than flying, when disturbed. Successfully established in Britain at the end of the eighteenth century, after several earlier attempts to introduce the 'Frenchman' had failed, the 'redleg' breeds in a number of localities in the South East. Breeding season counts of both partridge species would be welcomed by local ornithological societies.

Quail, which look rather like tiny partridges, have visited the three counties of our region in small numbers in recent years. They are very difficult to observe but have been heard calling from barley in Surrey and from the Downs behind Brighton. The only long-distance migrant of our game-birds, it is certainly worth listening for its characteristic 'wet-my-lips' call.

Water-fowl

Turning to water-fowl, the bird-watcher will find that the South East, with its varied stretches of water, marshes and coast-line, attracts many species.

Our resident, semi-domesticated mute swan, which is distinguished from Bewick's and whooper swans, two winter visitors from the north, by its orange bill with black base and knob, needs little introduction. It is commonly seen on lakes and rivers in towns, but breeding places include such other sites as ponds, marshes and flooded gravel workings. The main concentration of mute swans in Surrey is on the Kingston reach of the Thames. In winter large herds gather at places like Burham, Sarre marshes, High Halstow, and Walland Marsh, where 200 have been seen feeding on wheat. In severe weather immigrant mute swans swell the numbers reaching the South East as part of a general hard weather movement originating in Sweden and the Baltic, a fact discovered by recoveries of birds ringed during the hard winter of 1962-3.

Often ignored by bird-watchers, the mute swan is the subject of complaints from time to time. When it leaves the water, where it grazes largely on plant-life, to feed on land it can annoy the farmer. One can understand concern over abnormal feeding such as that noted during the 1962–3 cold spell, when mute swans fed on kale standing above deep snow and on cooked meat intended for pigs, because, as Dr Jeffrey Harrison has pointed out in *A Wealth of Wildfowl*, there is the possibility of abnormal feeding habits leading to an acquired taste! When they take to the air mute swans sometimes kill themselves and cause power failures by flying into overhead electricity cables, and there have been cases of these having to be moved to avoid regular flight lines used by the birds.

Many species including swans have benefitted from the magnificent joint efforts of Dr Jeffery Harrison and his father, the Wildfowlers' Association of Great Britain (WAGBI), the Wildfowl Trust and the Kent Sand and Ballast Company who allowed a reserve to be established on their flooded gravel pit at Sevenoaks in 1956. Thousands of trees and plants have been planted there, providing both food and cover.

Free-winged greylag geese were introduced at this WAGBI/Wildfowl Trust Experimental Reserve, and the first nesting success took place in 1964, when a pair of greylags from the reserve reared six young on a wooded island in a lake several miles away. All eight birds returned to the reserve in autumn and the following summer the parents reared three more goslings at their original nesting site. There is now thought to be an interchange of greylags between the West Kent gravel pits at Sevenoaks, Sundridge, Chipstead and Leigh. Also in Kent, bird-watchers have seen greylags with pinkish bills, which suggests that continental eastern greylag geese occasionally come as winter visitors to the South East, as resident greylags are typically orange-billed.

Our region attracts three other types of grey geese as winter visitors. White-fronted geese are seen flying overhead or feeding in the Thames estuary and in such other places as the Medway estuary, Stour valley and Norman's Bay, and small numbers of bean and pink-footed geese sometimes occur, too.

Two black geese visit our area in winter, the brent and the

much less numerous barnacle. The brent, our smallest British goose, makes its main winter quarters on the Thames, Swale and Medway, and in Chichester and Pagham harbours. An Arctic nesting species, the brent's favourite food is eel-grass or grass-wrack (*Zostera*), a submerged marine plant, but green seaweeds and certain grasses are also taken.

Representing the black geese as a breeding resident in the South East is the Canada goose. First introduced into Britain from North America some three hundred years ago, this large and fast-flying bird is now established wild here. Complaints have been made of damage to growing corn in summer, but the Canada goose's main foods are grasses, water plants, certain seaweeds and such forms of animal life as insects and worms. After nesting on lakes, ponds and flooded gravel workings, particularly those where there are natural or artificial raft islands, Canada geese gather in flocks, one such post-breeding concentration having reached almost 300 in the Sevenoaks area in late September.

Due perhaps to indecision on their real status, this and other introduced species have often been ignored with the result that their histories are by no means complete. Canada geese introduced into Sweden have taken to migrating south, to winter in Holland, a fact suggesting that this is a species whose habits and movements are well worth watching. There are, indeed, many opportunities for potential goose specialists in the South East, heavily populated and built-up as it is!

Likewise there is no shortage of topics for investigation by naturalists prepared to study the ducks of our region. County bird societies would, for instance, welcome breeding records of the shelduck, which may be spreading inland as a breeding bird, a development that could lead to new feeding habits. A striking goose-like duck that is increasing, it breeds along the coast at such places as Chichester and Pagham harbours and in the Rye area, while pairs are seen with young at certain gravel pits and other sites inland.

Often selecting a rabbit burrow for its down-lined nest, the shelduck will also lay its cream-white eggs under bushes and other cover. The young fly about eight weeks after hatching. In July they are deserted by their parents whose moult-migration enables the majority of British shelduck to spend the flightless moult period safely on sandbanks in Heligoland Bight. After the moult shelduck return to winter in parties along the coast and occasionally inland. Although there are times when it may include nereid worms, sea lettuce and seeds, the shelduck's diet consists almost entirely of the laver spire shell, a small and common marine snail of the mud flats.

A greater variety of food is eaten by our commonest British duck, the mallard. This ancestor of certain of our domestic ducks is very fond of grain, especially barley. Acorns are also taken, as are seeds of weeds and such trees as alder and birch. Animal foods include water snails, insects and their larvae. On brackish marshes seeds are eaten and so are brackish water shrimps and Jenkins's spire shells. Mallard have also been known to seize and swallow house sparrows and the stomach of one of these ducks contained 170 common shore crabs.

Mallard are no less adaptable in their choice of nesting site. Under normal conditions the turquoise-buff eggs are laid in a lined nest on marshy ground, often under sheltering bushes. But nests are made in hollow trees and in pollard oaks or willows as much as twenty-five feet from the ground. Near the Medway estuary mallard have nested in hay fields, where losses have occurred at haymaking. The mallard's natural enemies include foxes, rats, whose fondness for duck eggs is well known, and fish like carp and pike which devour many ducklings.

In autumn and winter, when birds may move in from other parts, large flocks of mallard congregate on estuaries, reservoirs and other large sheets of water in the South East, the breeding birds dispersing to marshes, gravel pits and small waters early in spring. Here the drake, with his beautiful metallic green head and neck and prominent white neck-ring, guards the nesting territory while the duck incubates. When dry the ducklings are taken to water and when seven to eight weeks old they can fly.

Each year these wild-bred birds are joined by hand-reared mallard released by wildfowlers.

These sportsmen have also liberated hand-reared gadwall, notably at the WAGBI/Wildfowl Trust Experimental Reserve at Sevenoaks. With its characteristic black-and-white wing-mark and white belly, the gadwall is increasing in Kent and has bred at Sevenoaks and on the Cliffe peninsula. This species does not appear to breed in Surrey or Sussex, but it may be seen in autumn and winter at Barn Elms Reservoir and Chichester gravel pits.

Other surface-feeding ducks known to breed in at least two of the counties of our region are the shoveler, readily known by its large, heavy spoon-shaped bill, the teal, our smallest duck, and the garganey, a summer visitor that winters in Africa.

Wintering numbers of wigeon are often high in coastal areas of Kent and Sussex, though few pairs remain throughout the breeding season. It is some time since breeding was confirmed in the region, but bird-watchers should be on the alert, though one hopes they will not emulate the naturalist who, in 1932, placed six eggs from a nest on the North Kent marshes in an incubator and showed that the three birds which hatched were wigeon.

The pintail, the species with the handsome long-tailed male, has bred in the South East, but the bird-watcher's best chance of seeing it is in winter when it gathers in places like Chichester harbour, the Medway estuary and the Thames marshes.

Although other diving ducks may be seen in the region, tufted duck and pochard are the commonest and the only ones that breed here. The tufted duck often associates with pochard in coastal areas and such inland freshwater localities as flooded gravel pits and reservoirs. Feeding largely on mollusca, crustacea and insects, with much smaller amounts of seeds and other plant foods, tufted duck are very fond of the zebra mussel, a freshwater species that has spread widely since it was first noticed in England in the Surrey Commercial Docks in 1824. By contrast, the pochard feeds mainly on pondweed seeds, stonewort, and other plant materials.

Passing mention must be made of the fish-eating saw-bills, namely smew, goosander and red-breasted merganser, which may be found on the coast or, in the case of the first two, on the London (Surrey) reservoirs.

A fish-eater that may be seen in many parts of our region throughout the year, the heron was once the quarry of royal hawks and a favourite item at great banquets. Nowadays this large, long-legged bird is left to go its own way and to feed in dykes, lakes and estuaries, occasionally leaving these to take goldfish from garden ponds.

Fishing herons may resemble motionless grey ghosts, but there is nothing still or silent about a heronry where nests, large swaying platforms of twigs, are built in tree-tops to hold the rough-shelled bluish-green eggs and later the young. In the South East the largest heronry is on the Thames estuary at Northward Hill, High Halstow, Kent, where more than one hundred pairs occupy nests. Here eels form the heron's most numerous prey, rudd being the only other important large fish that is taken. Large numbers of minnows, sticklebacks and marine shrimps are also eaten, as are a few water voles.

Feeding like this in shallow waters, herons can be literally frozen out of many of their normal haunts during a hard winter like that of 1963. In 1964 the number of occupied heron nests at Northward Hill was down to 80, but this had risen to 149 by 1968.

A heron-like bird that also suffers badly in severe weather is the bittern whose booming, or bog-bumping, is heard in early summer along the Stour valley of Kent, where at least one pair breeds. Despite legal protection of this rare bird and its eggs, there is no room for complacency over its safety. The shooting of a bittern at Reculver as recently as 1968 makes this painfully plain.

Continuous vigilance and energetic action are no less imperative where oil pollution is concerned. This is evident when, turning to review the grebes and divers of our region, one reads coastal observers' reports of black-throated and red-throated divers and great-crested, red-necked, Slavonian and black-necked grebes being killed by oil. What is needed is a public protest by naturalists and other sympathetic people every time evidence of such man-made misery and destruction is found.

The great-crested grebe has certainly suffered through oil in coastal localities in recent years, though it does not appear to have declined in the Medway estuary after 8,000 acres of saltings were polluted in September 1966, when a tanker accidentally lost some 1,700 tons of crude diesel oil while pumping her cargo ashore. Severely reduced by the plume-trade in England, Wales and Scotland to only about forty-two pairs in 1860, our largest British grebe increased with protection and now nests on gravel pits, lakes, reservoirs and similar large stretches of water. Here pairs adorned with tufts and tippets engage in complex courtship displays early in the year.

Though usually anchored to vegetation in the shallows, nests have been built on artificial raft islands. Normally four eggs are laid, but some bird-watchers are concerned about the great-crested grebe's apparently poor breeding success in parts of the region where, for example, twenty-five broods observed in Sussex averaged $1\frac{1}{2}$ young. They may find some consolation in the fact that, as long ago as 1924, Miss E. L. Turner found the proportion of nestlings to eggs to be comparatively small. This distinguished ornithologist spent many summers living in a houseboat situated between two nesting pairs of these beautiful birds with which, she considered, no species could compare for 'stateliness, noble bearing, and aristocratic exclusiveness'.

Feeding on fish, insects, small water creatures and plants, great-crested grebes are fine divers, staying under for 30–45 seconds. Their curiously striped young, which start to ride on the parent's back when only a few hours old, are delightful to see. Out of the breeding season great-crested grebes are commonly seen on the Channel and on tidal waters, on fresh water near the coast and at certain Surrey reservoirs.

Like its larger relative, the little grebe or dabchick is an expert diver. Also like the great-crested grebe, it covers the eggs in the heap of vegetation serving as nest whenever it leaves them. Feeding on small fish such as sticklebacks, tiny water creatures and some plants, the dabchick prefers undisturbed water where reeds provide cover. Displaced from some favourite haunts by mechanical dredgers, filling in of gravel pits, land drainage, and, as at one place in Surrey, the creation of cress beds, this dumpy little bird still breeds on ponds and gravel pits, and sometimes on marshes and the quieter reaches of rivers. Autumn and winter flocks occur mainly near the coast, though some can be seen on inland water.

At certain haunts of little grebes one may find the moorhen with its red forehead or the coot whose forehead is white. These two rails may, in fact, be seen on the same stretch of water, as at Winchelsea where, in early September, I have seen both species on a reed-fringed pond that is almost surrounded by bungalows.

Nesting throughout the region in suitable cover on ponds, lakes, gravel pits and cress beds, and by the slower rivers, the moorhen is less restricted in its breeding distribution than the coot, which usually selects larger stretches of water and does not often nest on rivers and streams. One reason for this is that the moorhen spends much more time on land than the coot. It is thoroughly at home there and always ready to supplement its diet of aquatic plants (especially duckweed), insects, small molluscs, worms and grass with bread, potatoes, fruit and scraps. It will feed with poultry and domesticated ducks, and is at home in urban areas. Many times I have passed a moorhen roosting in a bush above a stream beside a busy city road.

Moorhens may feed in flocks in winter, but coots make a regular practice of doing so, assembling in large numbers at places like Dungeness, Lydd, Chichester and Pagham. Arctic weather at the end of 1962 took a heavy toll of moorhens and coots.

Many water-rails also perished then. Looking like small dainty moorhens, they are shy and furtive in their habits, living in dense reeds and thick marsh vegetation where their harsh voices give them away. Normally feeding on seeds, roots of aquatic plants (those of watercress are a great favourite) and small water creatures, water-rails were seen during the 1962–3 cold spell attacking dunlins and knots which, like many other waders, were weak and reluctant to fly.

Waders

There were numerous casualties in 1962–3 amongst waders, particularly redshanks, but the bird-watcher will still find many members of this large and interesting group in the South East. Some may be seen throughout the year, but numbers and species vary considerably, especially as migrants pass through in spring and autumn and as flocks from inhospitable northern regions arrive to stay and winter.

Having sought expert advice as to local tide and weather conditions, the wader-watcher, warmly clad and water-proofed and provided with good binoculars, takes to estuaries and mudflats, sea-beaches and harbours. Here he watches oystercatchers, redshanks, curlews, sanderlings, dunlins, turnstones, and others, all seeking food and then, as feeding grounds are flooded by the rising tide, sees them flock for resting places. Inland, experienced watchers are not surprised to see waders such as snipe, curlews and green sandpipers turn up at reservoirs and sewage farms. Though some can be seen on estuaries and at the coast during the breeding season, many waders are at their breeding grounds then.

Unmistakable with its black-and-white plumage, long orange bill and pink legs, the oystercatcher nests on shingle at Dungeness and at several Sussex harbours, occasionally laying its brown-spotted stone-buff eggs in a 'scrape' in an adjoining field. Suffering from disturbance and egg thieves, the oystercatcher is often the victim of those who object to its fondness for cockles and mussels.

As a breeding species, the lapwing is far more widespread throughout the region than the oystercatcher, though it has taken some nine breeding seasons to recover from the halving of its population during the 1962–3 winter. With its long crest, broad blunt wings, and 'pee-wit' call, it is found nesting in water meadows and rough fields, on grass and arable, on chalk downland and at certain sewage farms. In Sussex, fourteen nests were found on a 3-acre field of kale, and one nest, a 'scrape' lined with stems, contained four eggs by 11 March. Like all baby waders, young lapwings are very active but their colouring affords great protection when they remain motionless. Feeding on insects at first, they later take worms, small molluscs, insect larvae and seeds.

With a somewhat similar diet, the redshank usually nests in water meadows and marshlands, though it has occasionally bred successfully in dry areas (including Goodwood motor-race track, Sussex, in 1966!). Its flute-like call is clear and beautiful.

Another sound that enlivens marshes in spring is the vibrant drumming of the common snipe. This occurs during flight as the bird makes a steep dive from a height and air rushes past and through two tail feathers held at right angles to the direction of flight. Improved drainage, with the consequent loss of marshy areas where the snipe can use its long straight bill to probe for worms, has undoubtedly contributed to the reduction of the breeding stock of this species.

Like the snipe, the woodcock, another long-billed wader, probes for worms in oozy patches. Rarely seen, it shelters among bushes and bracken in open woods, emerging at dawn and dusk to fly to its feeding grounds. It nests in the wood, using leaves to line a hollow on the ground. Probably more numerous in the region than is generally believed, the woodcock reveals its presence at dawn and dusk from early March to mid-July when the male goes roding, flying round above the tree-tops and uttering a shrill 'whick' and a peculiar croaking growl ('Like a frog with hiccoughs', as Kenneth Richmond so aptly put it).

Another sound that discloses its maker's whereabouts is the curlew's wild and nostalgic cry. This wader with a long, strongly down-curved bill is familiar at the coast, but very few pairs nest in the South East. Some are present in Ashdown Forest and on one of the Surrey commons during the breeding season, while north-west Sussex has another favoured site.

The curlew's precarious position as a breeding species in our region is shared by that long-legged wader with the large yellow eye, the stone curlew, the 'thick-knee' of my native Norfolk. A summer visitor to shingle banks at Dungeness and to the Downs, this shy bird with the weird wild cry is still pestered by egg collectors and the number of breeding pairs is very low. It would be a sad loss if the stone curlew, like the Kentish plover, now a very rare passage migrant, ceased to breed here.

This talk of a lost wader must not obscure the fact that, since the late thirties, the little-ringed plover has established itself as a British breeding species. It is distinguished from the common-ringed plover by the absence of white wing-bar, the presence of a narrow white head-band above the black forehead band, the dull flesh colour of the legs, and the higher pitched call-note 'pee-oo'.

Normally selecting flat, stony ground and areas of gravel not far from the shallow edges of freshwater, the little-ringed plover's nesting sites have included gravel pits, the gravelly bed of a reservoir under construction, a small stony patch at the end of an airport runway, and the top of clinker being dumped into an old gravel pit near Richmond. In the South East this summer visitor's breeding population is still small, especially in Sussex. It has lost some eggs to collectors and to carrion crows, but it it seems to be spreading.

Generally regarded as more a bird of the shingle beach, the common-ringed plover has suffered much in recent years from increasing public disturbance during the breeding season. Along parts of the Kent coast this has driven some pairs away from the foreshore to nest on gravel pits and workings nearby. Fortunately there are important nesting sites at Pagham harbour, Rye Harbour, and Dungeness, all places with official status as nature reserves.

Sea-birds

Like many other parts of the region, the three areas just mentioned attract numerous sea-birds. Several kinds of terns, all birds of beauty and grace, are on the regional list, but only two breed here.

The black tern, now seen as a passage migrant at the coast and on inland gravel pits and reservoirs, used to breed on Romney Marsh and more recently, in 1941, 1942 and possibly 1943, is reported to have bred at Pett Level, Sussex, which had been flooded. At the Ouse Washes (Cambridgeshire and Norfolk), where black terns have recently bred after a very long absence as a breeding species, water-covered areas, where nests can be built on floating vegetation, are favoured. Perhaps this insect-eating 'marsh' tern would be tempted back to the South East if suitable conditions could be established.

The region's two breeding terns nest in colonies, eggs being laid in simple or grass-lined scrapes in sand or shingle. Often close to one another, these nests are easily trodden on, and so considerate and unselfish bird-watchers keep off the breeding-grounds. Even from a distance terns are a delight to watch, especially when they are fishing.

One of our breeding species, the common tern, is more numerous than the other, the little tern, whose white forehead and yellow bill are distinctive features in the breeding season. Common terns have little patience with the chicks of neighbouring pairs and readily attack and even kill them. Though much less pugnacious than other terns, the little tern will make brave efforts to defend its young.

Terns may desert their nests if the weather suddenly becomes cold and wet. High tides can cause great losses of eggs and chicks, while blown sand may smother nests, sometimes burying the young. Predation in terneries can also cause severe losses. Stoats, weasels and rats can be a serious nuisance. In ten days a stoat accounted for sixty-one brooding terns, and in two nights a rat destroyed eighty clutches of tern eggs. Kestrels and short-eared owls will seize tern chicks, while carrion crows, gulls and oystercatchers destroy many clutches of tern eggs. Sadly, human disturbance and nest robbing must be added to this catalogue.

Adept at sucking the eggs and killing the chicks of common and little terns, and quite prepared to rob terns of fish to feed their own young, black-headed gulls can do much damage in a ternery and are therefore often discouraged. At Rye Harbour, where naturalists are keen to revive the common tern colony on Nook Beach, these gulls showed great persistence, remaking nests and laying further clutches after clearance of earlier ones.

Adaptability is another characteristic of this gull whose larger colonies are at Dungeness, Cliffe and Elmley in Kent, and at Rye Harbour in Sussex. It feeds at refuse tips, accepts bread and scraps, and, given the chance, follows the plough in search of earthworms and wireworms. On mudflats it tramples with both feet to bring up worms, while in the air it hawks gracefully for insects.

A non-breeding visitor to the South East, the great black-backed gull is a successful predator, killing and eating weak or wounded ducks and other birds. Lesser black-backed and herring gulls search marshes for eggs of ducks, lapwings and redshanks. A great scavenger, too, the lesser black-back breeds on cliffs east of Folkestone Warren and near Hastings. Always ready to devour stray chicks, herring gulls are sometimes controlled at bird reserves. They rear young on the roofs of houses at and near the coast, while the chalk and sandstone cliffs of Sussex held 716 occupied nests in 1969.

Likely to be confused with the herring gull, the common gull is best identified by its yellow-green bill and legs. Often consorting with black-headed gulls, common gulls are seen on farmland and playing fields in winter. Small colonies breed in the Dungeness–Lydd area, an interesting occurrence as most of our common gulls nest in north Britain.

The fulmar, unlike gulls, has tubular nostrils and is noted for its magnificent gliding flight. Feeding largely on oily matter, jelly-fish and other surface plankton, it has recently gained a foothold as a breeding species in Kent, where a few pairs have nested on cliff ledges between Folkestone and Ramsgate. Fulmars have been seen prospecting cliffs in Sussex and, before long, careful observers may well find their solitary eggs or chicks there.

Pigeons

Like the fulmar, the collared dove has extended its range dramatically. Indeed, its spread across Europe in the last forty years and its rapid colonisation of Britain since 1956, when it bred in Norfolk, are events of outstanding interest. Larger and greyer than the turtle dove, the collared dove, with its blackish primaries, lacks the creamy tints of the smaller Barbary dove, which occasionally escapes from captivity.

Now breeding in many parts of the South East, the collared dove likes to be near houses. Mainly a grain-eater, it finds food in poultry runs and farmyards and on land under stubble. Its nest, a mere platform of sticks, is sometimes started at the beginning of January and may be used for four or five broods, each of two young.

Unlike collared doves, which remain here all the year round, turtle doves are, as a rule, summer migrants, only one or two having wintered in Sussex. They nest where overgrown hedges, hawthorn and blackthorn bushes provide good cover.

Slighter than the wood pigeon and lacking its white wing-bar and neck-patch, the stock dove often feeds with its voracious relative whose flocks remain large, despite all attempts to control its numbers.

Crows

Sometimes grouped together as one 'superspecies', the all-black carrion crow and the hooded crow, a carrion crow with grey

Page 89 (above) Marshall's Lake, Bedgebury, Kent; important habitats of aquatic plants and creatures, such waters also have amenity value; *(below)* looking across the Weald from Mariner's Hill near Westerham, Kent; note the abundance of trees

Page 90   Birches, gorse and brambles flourish on coarse pebbly sand at Hayes Common, Kent

mantle and underparts, are both fond of eggs and small birds. Now a winter visitor and irregular passage migrant, the hooded crow has bred here. Usually nesting in trees, carrion crows have made use of a mobile crane in Kent, while in Sussex pairs have bred on an electricity pylon and a gantry.

A common breeding species throughout the region, the rook, whose nests are normally built in the tops of tall trees, nested on the framework of gasometers at Chichester for some years and even attempted to nest on cranes at Newhaven. The rook's long and varied list of food includes large insects, earthworms, acorns, walnuts and carrion, but it is the bird's liking for corn that often earns it a bad name, despite the fact that much of the grain is picked up among stubble.

Often associating with rooks in autumn and winter, jackdaws are common where suitable nesting sites are available in old and hollow trees, quarries and sandpits, or on sea cliffs and buildings. In belfries and chimneys their bulky nests of sticks can cause trouble for people involved. An omnivorous species, the jackdaw is unwelcome when raiding cherry orchards or taking young birds, but no one objects to its peculiar feeding-habit of delousing sheep.

Unlike jackdaws, magpies and jays are at home among bushes and in shrubby tangles. Though still shot in some areas, they are quite common, 'wild' woods and somewhat neglected gardens often acting as 'reservoirs' for them. The overgrown shrubbery of one garden I know at Chipstead, Surrey, attracts magpies, whose family parties are a delight to watch, while green woodpeckers visit the lawn and search for ants.

Other birds

Our largest British woodpecker, the green variety, is not always noticed, though its 'yaffle' or flight-note attracts attention. Now generally recovered from the effects of the severe winter of 1962-3 and again widespread, it seems to be less common than the great spotted woodpecker, which is not only well recorded

F

from wooded districts but known as a frequent visitor to gardens, even taking bread from bird tables. About the size of a sparrow, the lesser-spotted woodpecker may well be more plentiful than is often thought. Like its larger relatives, it should be sought where there are trees rotten enough for woodpeckers to excavate nesting-holes and to work for insect larvae.

Woodpeckers are sometimes plagued by that vigorous competitor for nesting-holes, the starling. One of the world's most successful birds, the starling is seen in most parts of our area. Adaptable and unspecialised in its habits, it breeds in woods and roofs. Truly omnivorous, its food includes grain, cherries, scraps, earthworms and slugs. A great mimic, it has an outstanding vocabulary. But perhaps the most fascinating of this smart and lively bird's activities are its fantastic aerial manoeuvres when groups returning at dusk from feeding grounds merge into vast pulsating clouds before descending to roost in some thicket or reed bed. One starling haunt south of Reigate is reported to have held some 250,000 birds in mid-September, and there are many other large roosts to be discovered in our region.

Like the starling, several other common birds visit gardens, which are one of the most important habitats for wild birds not only in the South East but in Britain as a whole. By making their gardens into bird reserves, by making drinking pools (line a hole with black polythene), providing nestboxes and bird-tables, and planting berry-bearing shrubs, naturalists not only benefit birds but bird-watching relatives and neighbours who are unable to get out and about much.

An RSPB research project produced an interesting list of plants with berries attractive to birds. Starlings took a particular delight in berries of elder (mainly *Sambucus nigra*) and holly (mainly *Ilex aquifolium*). Blackbirds favoured the berries of *Cotoneaster horizontalis* and hawthorns (mainly *Crataegus monogyna*), these latter shrubs also being popular winter roosts for large numbers of these birds. Song thrushes, which were watched feeding *Berberis* berries to fledged young at Woking in July, paid special attention to elder, yew (*Taxus baccata*), and *Cotoneaster horizontalis*. The familiar and successful house sparrow (about which many questions remain unanswered) took berries from at least eight different plant species. Additional food plants are blackberry, *Cotoneaster simonsii*, *Cotoneaster* × *watereri*, rowan, firethorn (*Pyracantha coccinea*), and crab apple. Even the berries of the oft-despised ivy are not neglected: blackbirds are taking them close to my window as I write in March.

Many names could be added to the berry-eaters. During really bitter spells the waxwing, a crested visitor from northern Europe, hunts hedges and gardens for berries and is usually quite tame. The fieldfare, a big thrush with conspicuous grey rump, and the redwing, a small thrush whose chestnut flanks and under-wing and buff-white eye-stripe make good recognition points, are other winter visitors from the north that are attracted by berries.

'God Almighty's Cock and Hen', the robin and the wren, will feed on berries. The robin does this much more than the wren, Kenneth Richmond's 'rusty imp', which has, however, been observed feeding on yew berries in September. Sadly ignored in many bird reports, Britain's national bird, the robin, is seen and heard in gardens, hedgerows, woods and parks all over the region. In the 'trough' of 1963 the wren lost a large part of its population but, happily, its loud bursts of song can again be heard in gardens, woods and hedges.

The great tit and the blue tit, both common species, eat several sorts of berries in winter, a time when peanuts, coconut, and other artificial supplies of food provided by thoughtful people help these and many other birds to survive. Originally woodland birds, most of our blue and great tits now live in gardens and farmland. In the South East blue tits can be quite enterprising in their choice of nesting sites, young having been reared in old wren's and house martin's nests and in the roofs of houses, while, like great tits, they make good use of holes in trees, walls and drain-pipes. In woodland, where the parents can collect the 700 caterpillars needed daily for their brood, the young that survive to become breeding birds represent only about one-tenth of the eggs laid. In gardens breeding success is often much poorer.

Our only small black-capped tit with a white patch on the nape, the coal tit is fairly common in woods and a frequent visitor to gardens, even quite small ones, in suburban areas. Broods have been raised in a hole in a wall in Uckfield High Street, in the roof of a house at Sanderstead, and in a wren's nest built in an old swallow's nest in a woodshed at Leigh in Kent. Marsh and willow tits are birds of woods and commons with scattered trees. Long-tailed tits, whose lichen-covered and feather-lined nests are things of beauty, leave their woods and bushy places to feed at bird-tables in winter at Haslemere and some other places.

Several finches frequent gardens where they are readily attracted to berries and sunflower seed-heads. A shy bird of wooded places, the hawfinch occasionally visits bird tables. Fond of hawthorn berries and beechmast, hawfinches have also been observed eating yew seeds discarded by thrushes on the North Downs in Surrey. Greenfinches come to gardens where I have seen them feeding on peanuts suspended in a net and on berries of *Daphne mezereum*, a shrub whose fragrant pink flowers appear early in the year. Goldfinches, birds of great charm that have increased much, breed in bushy places and gardens. Seed-heads of thistles and other members of the Compositae have a great appeal for them.

Looking rather like a miniature Colonel Blimp, as one writer put it, the bullfinch visits gardens, also being found in woods and on heathy commons where overgrown hedges and bramble tangles form excellent nesting sites. Living primarily on seeds, bullfinches turn to buds of hawthorn in woodland and those of fruit trees in cultivated areas when seeds become scarce. In the South East there has been severe damage to fruit buds in years when ash trees produced little or no seed. In an attempt to deal with this problem, fruit growers have destroyed large numbers of bullfinches, but the species continues to increase.

The chaffinch, a most abundant and widespread species, frequently visits bird tables in winter. Usually nesting in small trees or bushes, it will breed in large gardens, enlivening them with its bright call-note, 'pink, pink'. Even the brambling, a chaffinch-like winter visitor from Scandinavia, has visited bird tables, though it is essentially a ground-feeder commonly seen under beeches (as on the North Downs), often with chaffinches. Another seed-eating winter visitor, the siskin, feeds on alders and, less commonly, birches, sometimes in company with goldfinches and redpolls; the last-mentioned breed in bushy and wooded places and gardens. In such places the bird-watcher should not be surprised to see the tree creeper, a small brown bird with curved bill, moving mouse-like up (and occasionally down) tree trunks. Unlike the tree creeper, the nut-hatch, which creeps up and down tree trunks, is blue-grey above and has a straight bill. Nuthatches habitually nest in holes, and in Surrey one enterprising bird carried nesting material into a letter-box and plastered the lid with mud.

In bushy and wooded places and in larger suburban gardens, one sees the dunnock (or hedge sparrow). Brown above and grey on the underparts, this retiring little bird is streaked and spruce and certainly not 'plain and rather dull', as one bird report would have us believe. In the South East it is one of the chief fosterers of the cuckoo, others including meadow pipits, robins, reed warblers, pied wagtails, and whitethroats.

Arriving in the region in April (though occasionally very early birds can be found in March), the cuckoo calls its own name, thus making its presence known. The 'cuckoo's mate', the wryneck, once a familiar herald of spring, has gradually declined and the few pairs breeding in Kent and Surrey probably comprise the majority of English wrynecks. Naturalists should watch for it in large gardens—where it will use nesting-boxes—parks and orchards.

Many other summer visitors which come to Britain to breed are still seen in large numbers in the South East. Swifts are heard screaming as they rush through the air, feeding on minute in-sects. Less remote from man, swallows breed successfully in many places, their nesting sites including a fluorescent lamp and a rafter in a suburban garage. Besides nesting under eaves in country towns and villages, house martins have built in the chalk cliffs at Folkestone Warren. Many second broods are reared in the region, young still being fed in the nest as late as 13 October (1968) in Kent. Typically sand martins breed in colonies in sand, gravel and chalk pits and quarries, but in Surrey they have nested in drainage holes in a concrete river wall and in an actual river bank, while in Sussex use was made of drainpipes in a brick bank-facing on the Chichester canal.

Twittering and chirruping, swallows and martins do not, of course, possess voices as beautiful as those of some species. The nightingale's unique song, Milton's 'liquid notes that close the eye of day', is just as likely to be heard in daytime as at night in woods and thickets in many parts of the region. The blackcap, whose purity of voice is outstanding, is often regarded as a serious rival to the nightingale. With sudden snatches of high, flute-like notes, its song is distinct from the more sustained, lower-pitched warbling of the garden warbler. These songsters haunt woods and bushy places, as do willow warbler and chiff-chaff, those remarkably similar spring migrants.

A much less common species, the wood warbler breeds in open and mature woods of oak and beech and in wooded park-land. The grasshopper warbler, whose song has been described as like a free-wheeling bicycle, an angler's reel, or a distant lawn mower, is more likely to be heard than seen on marshes, commons and heaths. Sedge, reed, and marsh warblers occur in marshy places and by freshwater. An accomplished mimic and chatterer, the sedge warbler is known to be common and wide-spread, but further study is needed of the others, which have different songs but otherwise closely resemble one another.

Unlike these 'marsh' warblers whose throats are white, or almost white, the whitethroat and lesser whitethroat are not so attracted to damp places. Low tangled vegetation, including beds of nettles, attracts the whitethroat (or nettle-creeper), while the lesser whitethroat likes tall hedges and thorn thickets. An investigation into possible causes of the dramatic decrease in the British whitethroat population in 1968-9, which has certainly affected the South East, is being made.

Such an examination has already taken place in the case of the decline of the Dartford warbler, which is still seen in small numbers in a few places in our region and elsewhere in the south. Our only resident warbler, this small dark bird suffers badly during severe winters and springs when insects and spiders are hard to find. Nesting in gorse and heather and wintering in old, dense gorse, it falls victim to heath and gorse fires. The Dartford warbler will not (as Dr Norman Moore has shown) breed close to trees, which readily invade ungrazed and unmanaged heaths and commons. Add to all this the ever-present menace of the egg-collector and one applauds the efforts of the RSPB whose management of their reserve at Arne in Dorset provides for the building up of a good breeding stock of Dartford warblers from which other areas may be colonised.

Nesting populations of wheatears, small burrow-nesters, increased on the stony wastes of Dungeness after the RSPB warden placed empty ammunition tins in suitable sites, and one hopes that this will also happen at Rye Harbour reserve where land drain pipes are being put down to attract them. No longer grazed by sheep or rabbits, many areas have become too overgrown for wheatears and so, here again, habitat management (even in the comparatively simple ways mentioned) is vitally necessary.

We have come a long way since the days when Gilbert White and, as recently as 1900, W. H. Hudson were able to record the trapping of great numbers of wheatears, then esteemed a delicacy, on the South Downs. But it will be obvious to the reader who has followed me through this chapter that we cannot afford to be complacent and that the least we can do is to support the RSPB and the County Naturalists' Trusts.

Ringing and migration

There are bird observatories at Dungeness and Sandwich Bay, both of which issue annual reports and have certain facilities for visitors. The following table shows the numbers of birds ringed there.

	Birds ringed	No of species
Dungeness		
1952–67	87,444	169
1968	5,765	87
Grand total	93,209	171
Sandwich Bay		
1952–68	50,726	137
1969	2,470	68
Grand total	53,196	137

In addition there is a ringing station at Beachy Head, while individual ringers and groups operate in various parts of the region. Bird-watchers wishing to train as ringers or to assist with the work should make inquiries through county ornithological societies whose addresses can be obtained from reference libraries.

Recovery rates of ringed birds are often very low. For example, the recovery percentage for 11,525 Dungeness-ringed whitethroats (the species most ringed by that observatory) is 0·44, and this drops to 0·39 when 'local' recoveries are excluded. Some very interesting results have, however, been obtained. A redstart ringed at Dungeness in September 1966 was recovered in Morocco almost exactly two years later, and a whitethroat ringed by the same observatory in September 1967 was reported from Senegal (near Dakar) early the following year. Moorhens ringed near Dover were found dead or dying some two years later in Denmark and Germany, while goldfinches ringed at Beachy Head reached Northamptonshire, France and Spain.

CHAPTER FOUR

Trees and other plants

Beechwoods–Oakwoods–Birchwoods and pinewoods–
Heathlands–Chalk grassland–Chalk scrub–
Hedgerows, verges and ditches–Ponds–Fen, bog and
marsh–Salt marsh–Sand dunes–Shingle beaches–
Other places

ONE HAS ONLY to see the enthusiasm and hard work of amateur botanists, who are surveying and recording the flora of our region, to realise what a wealth of interest the South East still holds for those who are keen on plants.

In Sussex, the Flora Survey, originally envisaged as a ten-year project, has stimulated much activity, some 5,000 records being added during the three days of a Spring Bank Holiday meeting. In Surrey, 1966 saw the completion of fieldwork for a new Flora, during which the helpers' rewards included the discovery for the first time in the county of the slender sedge and the finding of several new localities for the common clubmoss, a species thought to be rare and decreasing in lowland areas. In Kent, too, much botanical work is in progress.

Certainly, those who appreciate the living beauty of trees, or enjoy the constantly changing woodland scene, will find much to please them in the South East. As I pointed out in Chapter One, though, they will discover that, owing to the complicated geological structure, the semi-natural woodland of the region varies considerably.

Beechwoods

On the chalk of the North and South Downs, beech, 'the Mother of the Forest', thrives. Foresters consider that, if grey squirrels can be kept away, it is the most suitable tree for the final crop here, though the plantations must, on silvicultural and economic grounds, contain conifers as temporary 'nurses' to the young beeches.

The types and numbers of smaller plants of beechwoods, or, to be more accurate, a particular part of a beechwood, depend on the amount of light penetrating the woodland canopy, the depth of soil, and the degree of moisture, to mention only a few factors.

In the real shade of close canopy beechwood, the naturalist should look for spurge laurel, an evergreen shrub whose faintly scented green flowers appear in early spring. Where there is plenty of decaying material in the soil, he may find saprophytes, plants which lack green leaves and are independent of light, such as bird's-nest orchid, with pale brownish flowers, and the yellowish or ivory-white bird's-nest. Toadstools of the genera *Cortinarius*, *Marasmius* and *Russula* are frequent on the woodland floor, while *Armillaria mucida*, with a very viscid, shining white cap, grows on the trees themselves.

Yew sometimes forms a second tree layer beneath beeches, as on some of the steep chalk slopes of Kent, but this evergreen is perhaps best seen at Kingley Vale, a National Nature Reserve where the largest and oldest yews include some that may be 500 years old.

There are some other fine yews in a number of south-eastern churchyards where the shining surface of the foliage lessens the somewhat sombre effect of its dark green tints. A few years ago, while attending a family wedding, I came across a wonderful old yew near the west end of St Peter's church at Tandridge, Surrey. On returning home, I discovered from Dr Vaughan Cornish's fascinating little book *The churchyard yew and immor-*

tality that this ancient specimen, with its immense trunk, was claimed to have the widest spread of branches of any yew in England (umbrage 8oft). Such a magnificent survivor reminds us that the yew, with its evergreen foliage, was long held to be a symbol of everlasting life.

Naturalists derive further interest from the fact that yew trees, both young and old, may bear 'artichoke galls' at the tips of shoots. Each gall, a tight cluster of 60–80 terminal leaves, encloses a single larva of the gall-causing midge *Taxomyia taxi* which pupates in the gall, the minute winged insect emerging a month later to carry on the life cycle of its species.

In autumn and winter some seventeen species of birds visit yews to feed on the bright crimson fruits, swallowing the large nut-like seed and its fleshy coating of sweet pink pulp, and often voiding the seed unharmed, sometimes far from the parent tree. Song thrushes and mistle thrushes are particularly fond of 'snotty gogs', as the berries are known in Sussex, but even wrens have been seen feeding on them. It has been suggested that birds must possess some immunity to the poisonous alkaloid taxine present in yew berries. Foliage, bark and seed are poisonous to man, and livestock may be affected, too, but rodents appear to be able to eat the kernel with impunity.

Another evergreen, the box, a shrub or small tree, grows in some south-eastern beechwoods, while a fine box-wood hangs on the west-facing slope of Box Hill in Surrey, which is now protected by the National Trust. Now scarce, box has inconspicuous green flowers, and in the mass smells strongly of foxes. However, in its role of host to a parasitic rust fungus, which affects the leaves, it is of interest to mycologists.

A deciduous tree seen among beeches, especially towards the woodland margin and where the canopy opens out, is the lovely whitebeam. This brings beauty to the woods as its white flowers appear in clusters and later when they give rise to bunches of edible scarlet fruit, and as the glistening white undersides of the leaves are revealed by the wind playing through the trees.

In similar places one may find several other small trees or

shrubs. The wayfaring tree, whose dense heads of cream-white flowers produce red fruit that eventually become black, is confined to lime-rich soils. This small tree or shrub has naked buds, while its leaves and twigs are covered in white down, a characteristic which, apart from helping to conserve water in really dry places, was responsible for the species' local names of cottoner and cotton-tree.

Buckthorn, another common shrub of chalk downs where beeches flourish, has small, green and inconspicuous flowers, and seems to lack significance now that its glossy dark black berries are so very rarely used as 'a mighty and muscular purge for men and cattle'. Nevertheless, it is a food-plant of larvae of the beautiful brimstone butterfly, a true harbinger of spring, and on its oval leaves one finds thickened marginal gall-rolls of the psyllid *Trichochermes walkeri* and yellow-orange aecidia of a rust-fungus whose alternate hosts are various grasses.

Recognised by its square twigs, the little spindle tree is a lime-lover with deep pink fruit whose opening reveals bright orange seed coverings. Like buckthorn, it plays a more important role in Nature than many people realise. Its greenish flowers are occasionally eaten by the yellow-warted larvae of holly blue butterflies, while its leaves are often rolled along the margins by mites. The spindle plays a vital part in the life history of the bean or black aphis, a common and widespread pest of beans and many other plants, both cultivated and wild, by serving as its winter host.

Dogwood, with blood-red twigs and crimson autumnal leaf tints, brings a touch of colour to areas of beechwoods, downland thickets and hedgerows. It sometimes serves as food-plant for larvae of holly blue and green hairstreak butterflies and a midge causes curious flask-shaped galls to project from the leaves.

Smaller beechwood plants growing in the less shaded areas include pink- or white-flowered wood sanicle, white helleborine, with flowers so aptly described by John Gilmour as 'looking rather like hard-boiled eggs garnished with green bracts', and drooping woodrush whose seeds are dispersed by ants. Where the soil is sufficiently deep and moist, dog's mercury often forms a 'carpet', spreading by a network of underground stems at the rate of six inches or more in a year. Fragrant-flowered sweet woodruff and wild strawberry may also appear in the ground flora.

Oakwoods

Dog's mercury and a few other plants already mentioned also grow in oakwoods, of which the South East has many fine examples, especially in the Weald. Both pedunculate oak and durmast (or sessile) oak occur here (for details of the characters of the two species see the *Biological Flora* account of *Quercus*, *J Ecol*, 47, 169–222). Pedunculate oak is dominant on moist, heavy soils, while the durmast oak is commonest on those that are lighter, more sandy and more acid.

Many of the oakwoods of the Weald Clay are noted for their under-storey of hazel. Though still used in this way, there were times when this catkin-bearing shrub was much more important as coppice, a crop cut on a regular rotation to provide rods for hurdle-making and for use in building. There is hardly a time when hazels do not help to support some other living thing and thus to provide something of added interest to naturalists. Early in the year bees that are fortunate enough to have hazel bushes near their hives will collect the light yellow powdery pollen so abundantly produced by the male flowers of the long 'lamb's tail' catkins and so rich in protein. Later insect larvae make patterns as they feed and tunnel in the leaves. Swollen leaf-buds, resembling the 'big-bud' of black currants, are found, these galls serving as living places for hundreds of mites, minute blind creatures that survive hard frosts but succumb to the direct rays of the sun. In summer, when the kernels are still milky and the shells far from hard, squirrels and birds like the nuthatch and great tit are already turning their attention to hazel nuts, an excellent source of energy-giving food. With the fall of leaves one may look more closely at the stems of hazel and at fallen branches and notice the presence of mosses, lichens and fungi.

The Wealden oakwoods may include trees such as birch, ash, field maple, wild cherry, holly, crab apple and hornbeam. Besides hazel there may be dogwood, blackthorn and other shrubs.

Putting matters at their simplest, the amount of cover formed by these trees and shrubs, together with the stage of development of the coppice, will help to determine the amount of light reaching the ground and thus the number and variety of smaller flowering plants. In oakwoods where the canopy is fairly open and two or three years have passed since the coppice was last cut, spring brings displays of wild flowers that are a joy to behold.

The bluebell makes food when the degree of light permits, storing it in bulbs, organs which provided our ancestors with glue and starch. Gathering its food into root tubers, the lesser celandine contributes golden yellow flowers to the spring display. The white-flowered wood anemone stores the food made during its short growth period in slender, creeping rhizomes.

These and other early spring flowers are followed by many more, such, for instance, as wood sanicle and wood avens, a species that, given the chance, interbreeds with water avens, producing 'hybrid swarms' of plants intermediate between the two parents.

Naturalists will find much else of interest in Wealden oakwoods. One such place, Nap Wood, 107 acres of oakwood in the Sussex Weald, includes one of the south-eastern habitats of the hay-scented buckler-fern, a plant whose distribution is predominantly western oceanic. Here, too, one finds mountain fern, large bittercress and marsh violet.

A number of the colourful spring and early summer flowers are absent from oakwoods on the lower greensand in Kent and Surrey, where durmast oak is dominant. Here there may be much bracken and many brambles, while such characteristic plants as wood sage and golden rod may also be found.

Bracken or 'fern' forms extensive colonies, spreading not only by means of long and slender underground rhizomes but also through the agency of microscopic wind-borne spores. The crozier-like young shoots may appear among bluebells early in the year, but they give rise to large green fronds whose cover and shade are often so complete that other plants find it very difficult, if not impossible, to gain a foothold. The dead, decaying fronds of bracken may make the ground unsuitable for colonisation by many other species. If permitted to do so, bracken will outgrow and smother young trees. It will also reduce the grazing value of pastures and has been accused of poisoning livestock. Like many other plants, bracken plays a part in the life of the insect world, and one finds on it leaf-mines, evidence of the feeding activities of minute flies, and 'little black puddings'—rolls along pinnule margins caused by tiny gall midges. In less affluent societies than our own this weed, pernicious as it is, has been used as litter for stock and even as a source of food for people.

Like bracken, brambles will spread in oakwoods and elsewhere, ousting less aggressive plants and making much work for those responsible for maintaining access to nature reserves and other places. Spreading freely by seed, often dispersed by creatures that have enjoyed the berries, brambles are assisted by their ability to produce root-shoots and the manner in which arching shoots so freely root at their tips. There is, however, no case for the general eradication of these thorny shrubs, despite the fact that, as we have seen, control may be considered to be necessary in certain fruit-growing areas where bullfinches are a nuisance. One has only to watch butterflies, bees and other insects feeding at the blossom, and birds and other creatures taking the fruit, to realise that a place must continue to be found for brambles, whose complex relationships and classification will continue to puzzle and occupy serious students of botany.

Among the numerous fungi of oakwoods, whose special species include several members of the genera *Boletus* and *Russula*, we may mention two of the larger ones. Beef-steak fungus, the *langue de boeuf* of the French, does, in fact, look rather like an ox-tongue growing from an oak trunk in late summer and autumn. Trees attacked by this fungus provide cabinet makers with 'brown oak', while adventurous spirits have been known to eat the fried plant itself. The large, bracket-shaped fruit-bodies of *Daedalea quercina*, a species which causes a reddish rot of oak wood, are common on old oak stumps, where they persist for a year or more.

Birchwoods and pinewoods

The naturalist who becomes fascinated by these and other fungi (a situation difficult to avoid once one has really begun to look at them) will find many more species in birchwoods and pine-woods, such as those on the sandy, acid soils of Surrey heaths and commons.

One 'pine-birch' species growing under these trees in autumn, the beautiful and conspicuous fly agaric, bears a shiny scarlet cap dotted with white patches. Despite its reputation as a killer, this species does not normally cause death in healthy people. But it should be avoided as it does give rise to delirium and hallucinations, sometimes accompanied by intestinal disturbances and always followed by intense stupor and an awakening to complete forgetfulness, a state of affairs which has led to three or four reindeer being paid for a single specimen in the Steppes. Common on birch trunks and branches, the razor-strop fungus is a bracket-like species whose dried-corky flesh was used for stropping razors.

Both birch and Scots pine regenerate naturally on south-eastern heaths, their pollen and seeds being readily dispersed by wind. Casting a fairly heavy and continuous shade, and shedding dead 'needles' that are slow to decay in the relatively dry conditions of our region, pines tend to discourage, and even prevent, the growth of many smaller plants. Small shrubby specimens of oak, beech and other broad-leaved trees survive for a time. Bilberry, the 'hurts' or 'whorts' of Surrey and Sussex, wavy hair-grass, and mosses such as *Dicranella heteromalla*, *Campylopus flexuosa*, and *Leucobryum glaucum*, are found in places where

Page 107 Birches on Holtye Common, Kent. Always ready to invade fresh territory, these beautiful trees are short-lived

Page 108 Wigeon jump from the saltings; wintering numbers of wigeon are often high in coastal areas, but few pairs remain

fallen pine-needles have yet to accumulate in depth. In light open spaces purple-flowered ling or heather flourishes, while cross-leaved heath, with rose-pink flowers, grows in wet heathy places.

'Ladies of the woods', birches are natural pioneers, invading open ground and filling clearings in woods. Light-demanding, they are often shaded out by more robust species (including pines). But, casting a light shade themselves, birches allow heath plants to grow under and among them when conditions are sufficiently open.

Here, as in other types of woods, competition between species brings inevitable changes in fauna, flora and scenery, whose progress is checked, and sometimes halted when man intervenes and fells trees or destroys vegetation, when rabbits are present, and when fires break out.

Heathlands

The naturalist is forced to admit, however, that without certain of the influences mentioned above, unpleasing as they may be at the time, even larger areas of heathland would be lost to invading pine, birch and scrub.

At the Thursley Commons Site of Special Scientific Interest the Surrey Naturalists' Trust, encouraged by the owners and helped by volunteer workers, has tackled this problem of conserving open heathland with informed enthusiasm and determination. Notified under the 1949 Act by the Nature Conservancy as an SSSI, the area includes almost the whole of Elstead, Thursley, Ockley, Royal and Bagmore commons.

Here, as on so many other commons and heathlands, the situation is far from simple. When commoners' rights fell into disuse, the consequent reduction in grazing animals severely weakened one control on the growth of vegetation. Invasion by pine started some 100 years ago, while military use of some parts in both World Wars also modified the vegetation. Plant life was again affected when vehicles were driven over the area, a prac-

G

tice that has been virtually stopped. Accidental burning brought changes, too, But in recent years fires have been fewer. Originally reduced by myxomatosis, the rabbit population is now officially controlled, activity that has necessitated not only the cutting of access rides through large areas of gorse, but the destruction of a pest which itself exerts a strong check on the growth of tree seedlings and other plants.

Intervention by the Trust has resulted in pine and birch being cleared to encourage the growth of ling or heather. Their efforts are also aimed at preventing encroachment by trees on the bog, a wet area where Sphagnum moss, orange-yellow-flowered bog asphodel and the insectivorous sundews and bladderwort are present.

As at Thursley site, heathland at Ashdown Forest has been subject to frequent fires. Indeed, hairy greenweed was destroyed there by fire in its last Sussex site. Some idea of the heath vegetation in this part of the region may be gained from a list of species growing at Gills Lap. In addition to Scots pine, ling and cross-leaved heath, there were bell heather, dwarf furze, common gorse, whose yellow flowers often brighten winter, and blue-flowered common milkwort. Also present were carnation-grass, a very narrow-leaved form of sheep's fescue, heath grass and purple moor-grass.

Ergot, a hard blackish, spur-shaped structure produced by the parasitic fungus *Claviceps*, is common on the last-named species, and it has been suggested that the presence of infected purple moor-grass may attract roe deer to the racing-rings where courtship activities occur at the period of the rut. Whether this is so remains a matter for speculation, but we do know that obstetrical use has been made of ergot for several hundreds of years and also that this fungus can be a potent poison. Modern research has shown that its chemistry is very complex and that its natural alkaloids act on the smooth muscles and the sympathetic nervous system.

The chemistry of the lichens of the genus *Cladonia*, which include the so-called 'reindeer mosses', 'cup mosses' and 'trumpet lichens', is no less complex. Several of these somewhat neglected plants occur at Gills Lap and in other parts of the region. One particularly attractive species, the red fruited *Cladonia floerkeana*, grows on earth with humus and on rotten wood and logs. It may contain, as an accessory substance, usnic acid. This compound is found in several lichens and forms the basis of antibiotics produced in Finland, Russia and Germany, where they are used in salves and also in conjunction with streptomycin in the treatment of tuberculosis.

Many other plants with beautiful flowers, fascinating chemistries or potential economic value await the naturalist on southeastern heaths, details of which are given in Chapter Seven, 'Places to visit'. Before leaving this inevitably brief section, however, attention must be called to the 'chalk heath', especially as the region has a fine example (some say the best example of its kind in Britain) at Lullington Heath on the South Downs. At this National Nature Reserve, a superficial acid layer of soil overlies deeper alkaline materials and so we see shallow-rooted, acid-loving ling and bell heather growing with such 'chalk species' as the deep-rooted salad burnet, with green, often purple-tinged, flowers, and dropwort whose cream-white petals may also be tinged reddish-purple.

Chalk grassland

At Lullington Heath, where the development of scrub has caused much concern in recent times (see p 42), grazing sheep and rabbits formerly helped to control the spread of trees, bushes and certain smaller plants, just as they did on the downlands generally. It is, in fact, true to say that the chalk downlands are basically a product of this grazing and that it, or a suitable substitute, is necessary in order to maintain their character.

Despite the decline in the number of grazing animals and the ploughing of much downland during and after the Second World War, there is still some fine chalk grassland along the escarpment of the North Downs in Kent and Surrey and on the South

Downs, particularly along the sea cliffs and the steep north-facing escarpment.

Obvious and successful downland plants, the grasses withstand both nibbling and wind and thrive on dry, thin chalky soils. Characteristic of these chalk soils, but not confined to them, are quaking grass, red fescue and sheep's fescue. Virtually exclusive to these highly calcareous soils are crested hair-grass, hairy oat-grass, meadow oat-grass, upright brome-grass and tor-grass, a locally dominant species that may cause a considerable risk of fire by producing a thick mat of dead vegetation and which may choke out many other plants.

The problems presented by tor-grass which is, as Sir Edward Salisbury called it, 'something of a chalkland weed', emphasise the importance of proper management of downland if many colourful and charming species are to survive. John Sankey, who reports having seen this grass effectively and drastically reduced by the grazing of a herd of Belted Galloway cows on the chalk scarps of the North Downs in Kent, is against burning it because its roots are barely affected by fire, which simply destroys less robust competitors and leaves more space for tor-grass to spread into. Pointing out that trampling by humans probably does as much as anything to keep it down, he suggests that it should be 'maltreated' as much as possible during the winter months.

Rough grassland dominated by tor-grass and upright brome-grass is not always entirely without interest. Among species that may be present are salad burnet and wild thyme, though it must be admitted that these herbs are often seen to best advantage elsewhere, thyme being generally more abundant in short sheep's fescue turf. Short downland turf, particularly that on southerly slopes, is also noted for purging flax, the small delicate herb formerly used as a purgative, and stemless thistle with prickly leaf-rosette and red-purple flower-head. Yellow-flowered horseshoe vetch, whose name came from the horseshoe-shaped segments of its curved seed-pods, thrives here, too, and merits attention and conservation not only for its own beauty of blossom and foliage, but because it serves as food-plant to two delightful butterflies, the chalkhill blue and the adonis blue. Vanilla-scented squinancy-wort, its quaint name recalling its ancient reputation as 'a sovereign remedy for the quincy', bears pink flowers on the chalk downs in June.

Of the numerous other wild flowers of south-eastern chalk grassland, the botanist will have no difficulty in finding small scabious, burnet saxifrage, bird's-foot trefoil (also called 'bacon and eggs' from the red-streaked yellow flowers), red clover, ribwort plantain and many more.

He is bound, sooner or later, to encounter one or other of the chalk orchids, the region, especially Kent, being an important stronghold for them. Observance of the code of conduct produced by the Botanical Society of the British Isles is imperative if these somewhat erratic and capricious plants are to continue to enrich our flora. Above all, information about the sites of rare and 'local' species should be restricted to local nature conservation trusts, which will try to help safeguard them.

As a rule, the typical downland orchids start to flower in May or June. Man orchid, its yellow lip shaped like a man's body, is one of the first to do so. A rather inconspicuous species, its smell is regarded by many people as unpleasant, but it may serve a useful purpose by helping to attract flies, ants and other small insects to the flowers. Man orchid is followed by the fragrant (or sweet-scented) orchid, a species sometimes found in large numbers and one that is visited by both moths and butterflies. Then comes the pyramidal orchid whose bright pink spikes do not appear for several years after leaves first form in the young plant's fifth year. Bee orchid, whose flower vividly resembles a bee visiting a blossom, is a downland plant whose roadside colonies in East Sussex have been given special protection by the Roads Department of the County Council. Also flowering in June and July, musk orchid is a rather dwarf honey-scented species whose favourite habitat is short turf. Autumn lady's tresses, another small orchid of low vegetation, bears small white almond-scented flowers in August and September.

Non-flowering plants of the chalk grassland include such

mosses as *Pseudoscleropodium purum*, *Rhytidiadelphus squarrosus* and *R. triquetrus* on damp northern slopes, and *Camptothecium lutescens* and *Fissidens cristatus* on drier southerly exposures. One of the best-known edible fungi, the St George's mushroom, so called because it first appears about St George's Day, 23 April, sometimes forms enormous rings on the Downs. A species producing an antibiotic substance that may become of use in medicine, it provides yet another example of a lowly plant which merits the attention of naturalists.

Chalk scrub

Quick to invade chalk grassland in the absence of grazing animals or human activity, scrub can have attractions of its own for the botanist. Hawthorn and juniper are early colonists on such ground. The latter has declined in the South East and the Sussex Trust is controlling hawthorn and other species vigorously at Saddlecombe Chalk Pit to prevent the juniper scrub, the last remaining example on the South Downs in East Sussex, from being overgrown. The Surrey Trust is engaged in similar scrub clearance from grassland round juniper bushes at Hackhurst, a National Trust property on the south slopes of the North Downs.

It would be a great pity if juniper, once widespread over Britain, were to disappear. A small coniferous bush, or miniature tree, whose blue-green, sharply pointed needles are grouped in threes, it emits a fragrant resinous odour. Its berry-like fruit, formerly swallowed to procure abortions and long used for flavouring gin, do not ripen until the second or third year and so one may observe their progress through green, blue and finally black on the same bush. A wind resistant species, juniper may help to shelter smaller plants in exposed places. It is attacked by a parasitic rust fungus, part of whose life cycle is spent on the leaves and fruit of the hawthorn.

Despite control measures against it and also the disappearance of hedgerows containing this vigorous shrub, hawthorn remains an essential part of the English countryside, the South East included, and one whose place in our folk lore is well established. Early in the year hawthorn brings fresh green to a countryside recovering from the austerity of winter, producing young leaves which people are still occasionally tempted to nibble, sometimes in complete ignorance of their old name of 'bread-and-cheese'. Later the leaves may be rolled into brightly coloured pouches by aphids and their margins may be rolled tightly by mites, while leaf-rosettes may form at the ends of shoots due to the activities of gall-midges.

In Sussex, where it is known as the cuckoo's bread-and-cheese tree, hawthorn is also called May, particularly when bearing its strong-smelling white blossom. These nectar-secreting flowers are visited by many kinds of flies, bees and beetles, and give rise to deep red fruit, the haws whose Old English name of haga appears in several forms in herbals and old books about Surrey and Sussex. Hawthorn would be worth conserving, and even encouraging in suitable places, if only for the great value of this fruit to hungry birds, especially in hard winters.

Other shrubs of chalk scrub include dogwood, spindle, buckthorn and the wayfaring tree, species already mentioned in the section on beechwood plants, and privet, elder and field maple. These shrubs all help to support certain other living things, as one is reminded by the red pimple-galls of mites on field maple leaves, and the bees and other insect visitors on privet blossom. I am reminded, too, that I once found larvae of a new species of gall-midge in unopened flower buds of privet. But it is the elder, so often uprooted, so frequently maligned, that I feel impelled to urge conservationists not to deny a place in nature reserves and the countryside generally. For it plays in its own way a useful part in the life of the wild. Its trunk and branches are the habitat of mosses and lichens and the curious jew's-ear fungus, flabby and ear-like in wet weather, but hard and horny when dry. Mined by a small fly and with their edges rolled by even smaller mites, elder leaves appear early in the year. Later, in June and July, the large clusters of creamy-white flowers are visited by small flies, which pollinate them. In early autumn the ripe purple-black elderberries attract hordes of starlings, though there are usually enough for other birds. Avoided by rabbits, elders are often selected by badgers for use as scratching trees, their thick soft bark being heavily marked by the constant scraping of the animals' claws. Thus this widespread shrub absorbs damage which might otherwise be inflicted upon trees and shrubs of economic importance.

Tree saplings frequently gain a foothold among the shrubs of chalk scrub, and in autumn, when this common climber and indicator of calcareous soils is in fruit, traveller's joy makes obvious the reason for its other name, old man's beard, as it puts forth dense white masses of long silky hairs. As the shrubs and trees grow and the scrub becomes denser, allowing less light to reach the ground, some smaller plants die out, to be replaced by others, such as dog's mercury, bugle, ground ivy, deadly night-shade and fly orchid, whose flower-lip is often mistaken by visiting male burrowing wasps for a female of their own species, so much so, in fact, that pseudo-copulation takes place.

Away from the chalk, the naturalist will not only find several other kinds of scrub, woody vegetation in which shrubs are dominant, but will come to distinguish between dense 'thicket scrub' and loose scrub that is progressing towards woodland.

Hedgerows, verges and ditches

Artificial forms of scrub, planted hedgerows, are still found in many parts of the region. Commonly of well-trimmed haw-thorn, they acquire added interest for botanists when neglected so that other shrubs, small trees, and climbing and scrambling plants are allowed to compete for light and living space.

Wild roses enliven the wilder type of hedgerow, acting as host to a rust fungus whose bright orange powdery masses are seen in May and June. They also form the habitat of several gall-causing insects, midges whose galls take the form of leaf-pods and wasps that cause bright red pea galls, some smooth, others spiked, and the beautiful mossy bedeguar galls, still often called robin's pin-cushions.

Bedeguar galls are at their most attractive in September, after which the reddish brown tints soon begin to blacken. In addition to the actual gall-causing insects, many of these galls yield parasites that have preyed on the gall-causers and also inquilines whose presence does not seem to harm other species. The bedeguar parasites may be attacked by hyperparasites, while the tissues of the galls themselves are often affected by a parasitic rust fungus. Bedeguar galls are, indeed, well worth the serious attention of naturalists interested in the lives and relationships of the inhabitants of small communities.

Given the chance, honeysuckle climbs in hedgerows, its deliciously sweet-scented flowers attracting hover flies, long-tongued hawk-moths and occasionally humble bees. A foodplant of marsh fritillary and white admiral butterflies, honeysuckle may, as we have seen, play a small but significant part in the life of the dormouse (see Chapter Two). In September its crowded clusters of red berries attract blackcaps and other small migrants, bullfinches, and marsh tits whose hovering attacks on the sweet and sticky fruits lend them the appearance of humming-birds.

White bryony and black bryony, with their poisonous red berries, are features of many hedges. Observations on black bryony have shown that forty-eight species of insects visit its green flowers, which last for $2\frac{1}{2}$–$4\frac{1}{2}$ days. Apart from thrushes and chaffinches, few birds appear to sample its berries. They are probably repelled by the presence of irritating raphides, minute needle-like crystals, and a poisonous saponin which becomes less abundant once the berries have been exposed to the weather. Water is stored in the deeply buried tubers of black bryony, and this assists its survival in time of drought, as does the presence of mucilage in the plant and of special hairs which burst and cover the surface of young parts with slime.

The semi-shade conditions of hedge-banks are favoured by cow parsley or Queen Anne's lace, rough chervil and hedge parsley, Geoffrey Grigson's 'roadside lace of high summer',

white-flowered umbellifers flowering in that order. Among
other flowers of this habitat are bush vetch, greater stitchwort
and lords-and-ladies.

Lords-and-ladies, a very common species of hedgerows and
woodlands, is a particularly fascinating plant, so much so that
Dr Cecil Prime devoted an entire *New Naturalist* volume, *Lords
and Ladies*, to it and its close allies. Judging by the large number
of local names, our ancestors must also have found it both
curious and amusing. I have collected fourteen such names from
our region, the longest being the Kentish 'Kitty-come-down-
the-lane-jump-up-and-kiss-me'. This and the others lack the
down-to-earth vulgarity of many of the names used elsewhere,
though the old name of 'cuckoo-pint' (Old English pintle or
pintel, the penis) plainly refers to the spadix, the long fleshy axis
on which the flowers are arranged, that stands enclosed in a
large sheath or spathe. The names 'lords-and-ladies' and 'ladies-
and-gentlemen' (Kent) refer to the barren upper part of the
spadix which is usually bright purple (lords), but sometimes
butter-yellow (ladies).

Simple flowers develop towards the base of the spadix, the
crowded males above the females. One particular species of
minute fly is attracted to these flowers, partly by a peculiar de-
composing smell, trapped by downwardly directed stiff hairs
for one night, and then, covered with pollen, released by the
withering hairs to visit other flowers and pollinate them. The
result of this activity is seen when the plant withers, leaving
what Audrey Wynne Hatfield once called 'a steeple of bright
scarlet berries', an early show of autumn colour. Though eaten
by pheasants, chaffinches and other birds, these berries are
reported to have caused the death of children.

Verges between hedgerows and roads may form important
relics of various types of semi-natural grassland and often merit
careful conservation. The commonest species of one such habitat
on the Bagshot Sand in Surrey were creeping soft grass, cock's-
foot and oat-grass, these being but three of the six grasses
present. Brambles and goosegrass also grew there, as did violet-

flowered ground ivy and the pleasantly aromatic herbs yarrow and mugwort.

Naturalists prepared to specialise on the plants of verges, which vary according to soil, drainage and other conditions, would gain much pleasure and information and be in a strong position to assist in the conservation of rare and interesting species. The same would be true for those willing to investigate roadside and hedgeside ditches whose flora is greatly affected by human interference and the amount of free water.

Ponds

Those who have helped with the survey of Surrey ponds, a project initiated early in 1965 by the County Naturalists' Trust, have done much useful work. Concerned about the gradual disappearance of ponds because of silting and subsequent invasion by land plants, filling in and other causes, they have compiled record cards for many ponds and a number of lakes.

It is not always appreciated that, during the course of natural development of vegetation, slow and gradual as this may be, a small isolated pond may fill up, or that the maintenance of such an area as a pond involves clearing it out every few years. The naturalist who finds a pond that has been left alone for several years should have no great difficulty in detecting the natural zonation and succession of plant communities.

The deeper water is occupied by submerged plants such as pondweeds and Canadian waterweed, an introduced species whose ability to spread and choke ponds and other waters earned it the name of drain devil. Next to the submerged zone come plants with floating leaves. Broad-leaved pondweed is one of these, while another is yellow water-lily, its flowers rising out of the water and smelling of alcohol (hence its Sussex name, brandy bottle). Plants, such as common spike-rush, with upright shoots rising above the surface, form another belt, and then there are taller species such as bur-reed, its flowers crowded in globose heads, the male heads above the female. Reeds, reedmace and other reedswamp species encircle the pond and then meadowsweet and other marsh plants may be found.

The raising of the level of the pond bottom, by the accumulation of dead plant remains and silt, gradually reduces the depth of water and forces submerged plants to give way to species with floating leaves and then to reedswamp plants which, in their turn, are followed by marsh plants.

The study of pond plants is mainly undertaken during the growing season. For, although some keep their leaves and continue to grow slowly during milder periods, many water plants die down and survive winter in the form of rhizomes, tubers or turions, which are stored with food. Turions become detached from the parent plant and, after floating or resting on the mud throughout the winter, develop into new plants in the spring.

Fen, bog and marsh

Despite the fact that many larger freshwater habitats have been destroyed or modified by drainage, grazing and cultivation, the South East retains several such areas as nature reserves.

At Stodmarsh National Nature Reserve, in the Stour valley of Kent, is an expanse of shallow pools and swamps where the reed forms extensive beds. Under natural conditions the dead annual remains of this large grass may accumulate under water and form peat, which gradually builds up to the water surface, so that other plants can gain a foothold in this purely organic soil (with its typically alkaline reaction) and eventually give rise to some type of fen community.

In east Kent, at one of the Kent Trust for Nature Conservation Reserves (see p 203), what is probably the last relic of true fen in our region is drying out and being invaded by scrub, owing to the general lowering of the water table by land drainage. Here one finds the marsh fern and the coarse perennial saw sedge, whose habitat is usually in situations where the water table is above root level.

Like fen, bog has a peat soil but, unlike fen-peat, this has a markedly acid reaction, a condition that some plants find intolerable. Bog plants growing at Thursley Commons SSSI were mentioned in the heathlands section of this chapter, and the naturalist should note that other interesting bogs survive in the heathlands of west Surrey.

Rich in mosses and liverworts and other typical bog plants, the valley bogs at Hothfield Common Local Nature Reserve, Kent, are well developed and intact. Bog-moss (mostly *Sphagnum pulchrum*) bears *Cephaloziella myriantha* and several other liverworts of the genus *Cephalozia*, with *Calypogeia trichomanis* on its surface among bog asphodel. The moss *Hypnum stramineum* creeps over the bog-moss, while *H. fluitans* grows among marsh St John's wort and bog pondweed in the boggy pools, around whose margins are zones of liverworts.

Other mosses and liverworts grow in less acid parts of the bog with such flowers as bog pimpernel and forget-me-not. Conservation measures at Hothfield Common, where heather covers ridges between the bogs, are aimed at controlling the spread of bracken and the invasion of boggy areas by birch and sallow scrub.

Sphagnum bog occurs on such Sussex commons as those at Ambersham, Heyshott and Chailey and, in a region where this is becoming rare, naturalists should help whenever possible to conserve the remaining small patches.

Marsh contrasts sharply with bog in many ways. It also differs from fen in having an inorganic (mineral) soil. But, that said, one must admit that the vegetation of marsh and fen is very similar, most of the species being the same. In the South East the trouble is that so heavy have been the demands for grazing and agriculture generally, many areas referred to by this name are no longer strictly marshes. Botanists will, however, readily find such marsh flowers as purple loosestrife, water forget-me-not, kingcup or marsh marigold, great hairy willow-herb and meadow-sweet in wet places by rivers, streams and lakes and in ditches, while some, such as kingcup and meadow-sweet, also grow in wet woods.

Those who are prepared to travel in search of wild flowers, and are ready to withstand occasional disappointment, will, in time, see many more attractive species. Thus a visit to Amberley Wildbrooks, the Sussex Trust's reserve on the wide alluvial plain east of the Arun, might well be timed to coincide with the flowering of the common marsh orchid (also called 'fen orchid') whose lilac-pink or pale pink flower spikes make a fine sight from early June to mid-July.

This area, where drainage dykes border small fields, has a rich flora, including cut-grass, marsh cinquefoil and the insectivorous bladderwort. Kingcups abound at Balcombe Marsh, another Sussex reserve, and flowers of two orchids, the marsh helleborine and the fragrant orchid, are seen during July. Here self-sown alders (if they have not been removed by then) serve as reminders that marsh, like many other types of land, is naturally colonised by shrubs and trees.

Salt marsh

Unlike those of freshwater habitats, plants of salt marshes are adapted to the salt content of the sea water which covers them from time to time. In our region, such species are found along the Thames and Medway estuaries, at Chichester and Pagham Harbours, Cuckmere Haven, and certain other places on the coast and estuaries.

Glasswort or marsh samphire is a highly succulent plant with erect jointed stems that lack proper leaves. On some south-eastern sites this early coloniser of mud is associated with seablite, whose tissues may contain considerable amounts of salt, fleshy-leaved sea-aster, an attractive relation of the garden Michaelmas daisy, and sea manna-grass, the commonest salt-marsh grass.

At Cuckmere Haven the last two plants are accompanied by rice-grass, a hybrid between the American *Spartina alterniflora* and the European *S. maritima*, which can colonise deep tidal mud and is able to reclaim land from the sea and establish *Spartina* meadows where cattle may graze. Also present there are sea plantain and sea purslane, a small shrub whose leaves are silvery-white as a result of a mealy coating of fine white scales on both sides.

Sand dunes

Unlike salt marsh species, most plants of coastal sand dunes are not adapted to the effects of salt water in the soil, though they must tolerate some salt-laden spray. It is their ability constantly to push fresh shoots through wind-blown sand that often prevents them from being completely overwhelmed.

Despite the fact that much of our south-eastern coastline has been, and continues to be, 'developed', sand dunes survive in a few places, such as Camber Sands, Sussex, and Sandwich Bay, Kent. Confined to this habitat, marram grass has stems that creep and root and bind the sand, helping to build up the dunes. Like this sturdy grass, sea holly, whose small blue flowers are in tightly-packed globular heads, is capable of vigorous upward growth when buried in loose sand. Its leathery bluish-green leaves have a tough spiny edge and a thick waxy cuticle, while its creeping roots may be as much as eight feet in length. Stabilising sand by means of its long spreading roots, sea bindweed sports beautiful pink flowers, parts eaten by rabbits which do not touch the leaves.

Many more flowers and several mosses have been recorded from Camber Sands where sea buckthorn grows on fixed and partially fixed sand. This prickly shrub stabilises sand, helping to prevent erosion, and has been planted for this purpose along several parts of the coast, where it is also used to screen unsightly buildings and defence works. It has been planted in gardens and roadways for the amenity value of its silvery foliage and orange berries. Despite reports to the contrary, the fruit is eaten by birds, the seeds having been found to remain viable after being ejected by hooded crows as pellets and after passing through the bodies of starlings. The fact that its fruit is exceptionally rich in vitamin C led to sea buckthorn being protected in Germany during the last war and to plant breeding studies concerning fruit production being carried out since then.

Shingle beaches

In the coastal part of the region, between Winchelsea and Hythe, are the largest expanses of shingle in Britain. Consisting largely of water-worn pebbles, this is the habitat of a number of interesting species, several of which were seen with both flowers and fruit in the Rye Harbour Local Nature Reserve area in early September. Yellow-horned poppy bore deep yellow flowers and long curved seed-pods, while the purplish-red blossoms of sea pea were accompanied by straight pods. Sea bittersweet, a prostrate form of the common bittersweet or woody nightshade, was well supplied with blue flowers and red berries, and bird's-foot trefoil had red-streaked yellow flowers. The large cabbage-like sea-kale bore egg-shaped seed-pods. More floral colour was provided by the blue of viper's bugloss and the red of red valerian, a showy introduced species.

Known in both Kent and Sussex as 'pretty Betsy' and in the first-named county as 'sweet Betsy', red valerian is a lovely flower, a native of the Mediterranean region that is as much at home on old walls as on shingle, in railway cuttings as in the old chalk quarries of north Kent. There is a white-flowered variety, but the 'normal' form produces flowers of numerous shades of red, which are pollinated by long-tongued insects.

Many of the Rye shingle plants also grow at Dungeness on the borders of Kent and Sussex where, despite the construction of a nuclear power station, much remains to interest the naturalist. Once proposed, although not actually declared, as a National Nature Reserve, this area was considered by the Nature Conservancy, in evidence given at the public inquiry into the proposed nuclear power station in 1958, to be 'the most significant shingle foreland in the British Isles and Europe and one of the major coastal depositional features to be found anywhere in the world'.

Page 125 Mute swans; in severe weather, when large herds may be seen, the resident, semi-domesticated birds are joined by immigrants

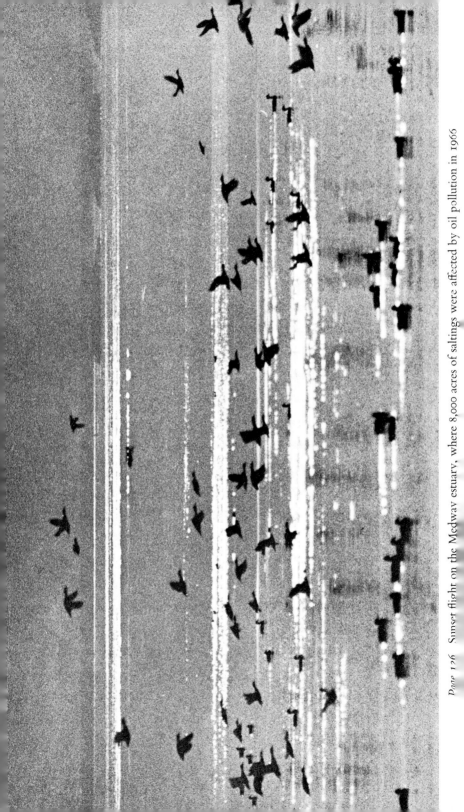

Page 126 Sunset flight on the Medway estuary, where 8,000 acres of saltings were affected by oil pollution in 1966

Showing much variety in its vegetation, the Dungeness area includes large tracts of bare shingle and other places where lichens alone grow on the pebbles themselves. Yellow-horned poppy, sea-kale and a grass, creeping fescue, colonise bare shingle. Lichens of the genus *Cladonia* and the mosses *Dicranum scoparium* and *Hypnum cupressiforme* var *tectorum* grow in shingle 'lows'. As humus forms from mosses and lichens, sweet vernal grass and sheep's fescue move in, as do yellow-flowered mouse-ear hawkweed, English stonecrop, sheep's bit, foxglove, Nottingham catchfly and several others.

Mention of Nottingham catchfly as one of the botanical attractions of Dungeness must not be allowed to give the impression that this plant is confined to shingle. For in the South East it also grows on the exposed edges of chalk cliffs near Dover and on gravel near Richmond in Surrey, and it is recorded from walls and field borders, too. True, it seems to have done well at Dungeness, one of the very few places in Britain where seedlings have been seen around wild plants of the species. This happy state of affairs for the plant has existed despite the fact that a powdery smut fungus infected a large percentage of Nottingham catchfly flowers at Dungeness some years ago, and also despite the presence there of common dodder, a slender parasitic plant which attaches itself to the catchfly, its host, by means of suckers.

Later stages reached by vegetation at Dungeness include the development of mats of a prostrate form of common broom whose branches provide shelter for other species, and scrub of stunted blackthorn, gorse, elder and holly, their branches, in spite of their exposure, bearing mosses and lichens. The shingle over the site of the power station was rich in the low-growing broom. The loss of much of this plant was much more than the destruction of an interesting form of a fairly common species. For this broom is the food-plant of an ochreous yellow form of the grass eggar moth, a variety which is believed to be unique to Dungeness.

Close to the site of the nuclear power station are the Open

H

Pits or Hoppen Pits. The only freshwater lagoons of any size over shingle beach anywhere in Britain, they are surrounded by neutral marsh vegetation, including marsh cinquefoil and marsh fern. Mosses grow there and also on damp shingle round this marsh.

Other places

Even for those who may find difficulty in visiting such places as the 'wind-swept wastes' of Dungeness, the Downs or the salt-marshes, the South East is not without botanical interest. Wild plants are widespread organisms and towns, villages and suburbs have numerous species on waste ground, refuse tips, and golf courses and in garden borders, churchyards and many other places. Common as some of them may be, they should be carefully identified and recorded in the card indexes and journals of local natural history societies.

The *Atlas of the British Flora* and its *Critical Supplement* will show whether a flowering plant or fern has been recorded from a particular ten kilometre square of the national grid and will also enable the botanist to prepare lists for the region or its individual counties. The distribution of mosses, liverworts and lichens is summarised on a vice-county basis in census catalogues issued by the British Bryological and British Mycological Societies (see pp 201–2).

Scientific names of species mentioned in this chapter:

Alder, *Alnus glutinosa*
Ash, *Fraxinus excelsior*
Autumn lady's-tresses, *Spiranthes spiralis*
Beech, *Fagus sylvatica*
Beef-steak fungus, *Fistulina hepatica*
Bee orchid, *Ophrys apifera*
Bell heather, *Erica cinerea*
Bilberry, *Vaccinium myrtillus*
Birch, *Betula pendula*, *B. pubescens*

Bird's-foot trefoil, *Lotus corniculatus*
Bird's-nest, *Monotropa hypopitys*
Bird's-nest orchid, *Neottia nidus-avis*
Black bryony, *Tamus communis*
Blackthorn, *Prunus spinosa*
Bladderwort, *Utricularia minor, U. vulgaris*
Bluebell, *Endymion non-scriptus*
Bog asphodel, *Narthecium ossifragum*
Bog pimpernel, *Anagallis tenella*
Bog pondweed, *Potamogeton polygonifolius*
Box, *Buxus sempervirens*
Bracken, *Pteridium aquilinum*
Brambles, *Rubus* species
Brandy bottle, *Nuphar lutea*
Broad-leaved pondweed, *Potamogeton natans*
Broom (prostrate form), *Sarothamnus scoparius* ssp *prostratus*
Buckthorn, *Rhamnus cathartica*
Bugle, *Ajuga reptans*
Burnet saxifrage, *Pimpinella saxifraga*
Bur-reed, *Sparganium erectum*
Bush vetch, *Vicia sepium*
Canadian waterweed, *Elodea canadensis*
Carnation-grass, *Carex flacca*
Cock's foot, *Dactylis glomerata*
Common clubmoss, *Lycopodium clavatum*
Common gorse, *Ulex europaeus*
Common marsh orchid, *Dactylorchis praetermissa*
Common milkwort, *Polygala serpyllifolia*
Common spike-rush, *Eleocharis palustris*
Cottoner, *Viburnum lantana*
Cotton-tree, *Viburnum lantana*
Cow parsley, *Anthriscus sylvestris*
Crab apple, *Pyrus malus*
Creeping fescue, *Festuca rubra*
Creeping soft grass, *Holcus mollis*
Crested hair-grass, *Koeleria cristata*

Cross-leaved heath, *Erica tetralix*
Cuckoo-pint, *Arum maculatum*
Cuckoo's bread-and-cheese tree, *Crataegus*
Cup-mosses, *Cladonia* species
Cut-grass, *Leersia oryzoides*
Deadly nightshade, *Atropa bella-donna*
Dog's mercury, *Mercurialis perennis*
Dogwood, *Thelycrania sanguinea*
Drooping woodrush, *Luzula forsteri*
Dropwort, *Filipendula vulgaris*
Durmast oak, *Quercus petraea*
Dwarf furze, *Ulex minor*
Elder, *Sambucus nigra*
English stonecrop, *Sedum anglicum*
Fen orchid, *Dactylorchis praetermissa*
Field maple, *Acer campestre*
Fly agaric, *Amanita muscaria*
Fly orchid, *Ophrys insectifera*
Forget-me-not, *Myosotis secunda*
Foxglove, *Digitalis purpurea*
Fragrant orchid, *Gymnadenia conopsea*
Glasswort, *Salicornia* species
Golden rod, *Solidago virgaurea*
Goosegrass, *Galium aparine*
Greater stitchwort, *Stellaria holostea*
Great hairy willow-herb, *Epilobium hirsutum*
Ground ivy, *Glechoma hederacea*
Hairy greenweed, *Genista pilosa*
Hairy oat-grass, *Helictotrichon pubescens*
Hawthorn, *Crataegus monogyna*
Hay-scented buckler-fern, *Dryopteris aemula*
Hazel, *Corylus avellana*
Heather, *Calluna vulgaris*
Heath grass, *Sieglingia decumbens*
Heath rush, *Juncus squarrosus*
Hedge parsley, *Torilis japonica*

Holly, *Ilex aquifolium*
Honeysuckle, *Lonicera periclymenum*
Hornbeam, *Carpinus betulus*
Horseshoe vetch, *Hippocrepis comosa*
Jew's-ear fungus, *Auricularia auricula-judae*
Juniper, *Juniperus communis*
Kingcup, *Caltha palustris*
Kitty-come-down-the-lane-jump-up-and-kiss-me, *Arum maculatum*
Ladies and gentlemen, *Arum maculatum*
Large bittercress, *Cardamine amara*
Lesser celandine, *Ranunculus ficaria*
Ling, *Calluna vulgaris*
Lords-and-ladies, *Arum maculatum*
Man orchid, *Aceras anthropophorum*
Marram grass, *Ammophila arenaria*
Marsh cinquefoil, *Potentilla palustris*
Marsh fern, *Thelypteris palustris*
Marsh helleborine, *Epipactis palustris*
Marsh marigold, *Caltha palustris*
Marsh St John's wort, *Hypericum elodes*
Marsh samphire, *Salicornia* species
Marsh violet, *Viola palustris*
May, *Crataegus*
Meadow oat-grass, *Helictotrichon pratense*
Meadow-sweet, *Filipendula ulmaria*
Mountain fern, *Thelypteris limbosperma*
Mouse-ear hawkweed, *Hieracium pilosella*
Mugwort, *Artemisia vulgaris*
Musk orchid, *Herminium monorchis*
Nottingham catchfly, *Silene nutans*
Oat-grass, *Arrhenatherum elatius*
Pedunculate oak, *Quercus robur*
Pondweeds, *Potamogeton* species
Pretty Betsy, *Centranthus ruber*
Privet, *Ligustrum vulgare*

Purging flax, *Linum catharticum*
Purple loosestrife, *Lythrum salicaria*
Purple moor-grass, *Molinia caerulea*
Pyramidal orchid, *Anacamptis pyramidalis*
Quaking grass, *Briza media*
Queen Anne's lace, *Anthriscus sylvestris*
Razor-strop fungus, *Polyporus betulinus*
Red clover, *Trifolium pratense*
Red fescue, *Festuca rubra*
Red valerian, *Centranthus ruber*
Reed, *Phragmites communis*
Reedmace, *Typha*
Reindeer-mosses, *Cladonia* species
Ribwort plantain, *Plantago lanceolata*
Rice-grass, *Spartina townsendii*
Rough chervil, *Chaerophyllum temulentum*
St George's mushroom, *Tricholoma gambosum*
Salad burnet, *Poterium sanguisorba*
Sallow, *Salix*
Saw sedge, *Cladium mariscus*
Scots pine, *Pinus sylvestris*
Sea-aster, *Aster tripolium*
Sea bindweed, *Calystegia soldanella*
Sea-bittersweet, *Solanum dulcamara* var *marinum*
Seablite, *Suaeda maritima*
Sea buckthorn, *Hippophae rhamnoides*
Sea holly, *Eryngium maritimum*
Sea-kale, *Crambe maritima*
Sea manna-grass, *Puccinellia maritima*
Sea pea, *Lathyrus japonicus*
Sea plantain, *Plantago maritima*
Sea purslane, *Halimione portulacoides*
Sessile oak, *Quercus petraea*
Sheep's bit, *Jasione montana*
Sheep's fescue, *Festuca ovina, F. ovina* ssp *tenuifolia*
Slender sedge, *Carex lasiocarpa*

Small scabious, *Scabiosa columbaria*
Spindle, *Euonymus europaeus*
Spurge laurel, *Daphne laureola*
Squinancy-wort, *Asperula cynanchica*
Stemless thistle, *Cirsium acaulon*
Sundews, *Drosera* species
Sweet Betsy, *Centranthus ruber*
Sweet-scented orchid, *Gymnadenia conopsea*
Sweet vernal grass, *Anthoxanthum odoratum*
Sweet woodruff, *Galium odoratum*
Tor-grass, *Brachypodium pinnatum*
Traveller's joy, *Clematis vitalba*
Trumpet lichens, *Cladonia* species
Upright brome-grass, *Zerna erecta*
Viper's bugloss, *Echium vulgare*
Water avens, *Geum rivale*
Water forget-me-not, *Myosotis* species
Wavy hair-grass, *Deschampsia flexuosa*
Wayfaring tree, *Viburnum lantana*
Whitebeam, *Sorbus aria*
White bryony, *Bryonia dioica*
White helleborine, *Cephalanthera damasonium*
Wild cherry, *Prunus avium*
Wild rose, *Rosa* species
Wild strawberry, *Fragaria vesca*
Wild thyme, *Thymus drucei, T. pulegioides*
Wood anemone, *Anemone nemorosa*
Wood avens, *Geum urbanum*
Wood sage, *Teucrium scorodonia*
Wood sanicle, *Sanicula europaea*
Yarrow, *Achillea millefolium*
Yellow-horned poppy, *Glaucium flavum*
Yellow water-lily, *Nuphar lutea*
Yew, *Taxus baccata*

CHAPTER FIVE

Freshwater fishes

Lampreys–Herring-like fishes–Salmon, trout and
grayling–Smelt–Pike–The carp family–Stone loach–
Eel–Sticklebacks–Bass–Perch and ruffe–Grey mullets
–Bullhead–Flounder

IN THE SOUTH EAST, as elsewhere, fish are all too often 'out of sight and out of mind', even where many experienced naturalists are concerned. This is a great pity, for they are creatures of considerable interest whose distribution has, in several cases, still to be worked out in detail.

Fish are taken as food by man, otters and certain other mammals, and by such birds as herons, kingfishers and gulls. The pike preys on water-birds, small fish, and mammals like the water shrew, while many kinds of fish take a heavy toll of insects and other invertebrate animals. And, of course, we must not overlook the fact that pollution of estuaries and rivers may not only affect fish but mammals, birds and insect life, too.

Fish are such a vital part of the life of rivers, streams and lakes that more naturalists should co-operate with Fishery Officers and anglers so that useful information is properly recorded. Published by the Freshwater Biological Association in 1972, Dr Peter Maitland's *Key to British Freshwater Fishes*, with its notes on distribution and ecology and its fifty-five distribution maps, makes a convenient starting point for anyone interested in taking up these studies, and the names and order of species adopted in that work are followed here.

Lampreys

Slimy-skinned and eel-like, lampreys are primitive creatures lacking bones and true jaws. Each of the three species found in northern Europe has been recorded from our region, but Maitland does not include the sea lamprey in the appropriate part of his distribution map. Perhaps the blind, toothless larvae have escaped detection in recent years. After spending 5–6 years buried in the mud of backwaters, they undergo metamorphosis and descend to the sea where they attach themselves to various fishes, sucking out their hosts' blood. Later they return to fresh water to breed.

Maitland records the river lamprey (or lampern) from one place in the region, the Thames, at the northern boundary of Kent, but this species also occurs in Sussex. Like those of the sea lamprey, the young (known as prides) live buried in rich mud. After six years the newly metamorphosed adults migrate to the sea where they attack fishes and spend part of their adult life before ascending rivers to spawn and die.

Unlike its relatives, the brook lamprey remains in fresh water throughout its life. The larval period in the mud lasts for five years and then the new adults migrate upstream where, without feeding, they spawn and die. This species is common and widespread in tributaries in Sussex.

Herring-like fishes

According to the *Victoria County Histories* (1902–8) the allis and twaite shads, the so-called 'king of the herrings', used to visit south-eastern rivers, having been found in the Thames as far as Hampton Court. Maitland's maps do not include the allis shad for the region, but they show a post-1960 record for East Sussex for the more widespread and common twaite shad. Sea anglers still take shad off such places as Deal and Dover, but man-made obstacles and pollution undoubtedly prevent many of these fish from ascending our estuaries and rivers to spawn.

Salmon, trout and grayling

That swift and powerful swimmer, the salmon, is famous for its ability to overcome such obstacles as weirs, rapids and small waterfalls. It was largely due to industrialisation and pollution that this 'king of freshwater fish' deserted so many of our south-eastern rivers. In 1681 James Chetham, author of *The Fly-fisher's Vademecum*, regarded the Thames as of 'principal note' as a salmon river, and even in the period 1794–1821 as many as 7,346lb of salmon were taken from its waters. The last recorded capture of a Thames salmon was in June 1833. Some thirty years later, when these fish were still trying to ascend the Thames estuary, attempts were made to restock the river with salmon. For several years thousands of fry were annually put into the Thames at Sunbury and Molesey. In 1873 more were introduced at Staines, Shepperton and Windsor. Then, between 1901 and 1906, some hundred thousand salmon smolts and parr were released in the river, mostly at Teddington. Sadly, all these attempts failed and no mature fish was taken.

In 1866 Frank Buckland's efforts to make the Stour a salmon and trout stream resulted in the removal of weirs and stake-nets, the erection of fish-passes, and the release of young fish into the river. It has been stated that, while the trout flourished, pollution prevented the development of the salmon fry. Happily, I am able to report that salmon are now occasionally found in the river Stour near Canterbury. Perhaps we may look forward to the day when, pollution eliminated, many more salmon will enter this and other south-eastern rivers, make their 'run' to the spawning grounds and lay their orange eggs in redds among the gravel before returning to the sea.

Similar to the salmon in certain aspects of its appearance and habits, the sea trout is, like the brown trout, a form of *Salmo trutta*, a species which varies in colour and body shape according to its surroundings and habits. The sea trout runs into a number of our south-eastern rivers where it breeds between October and January. They include the Rother and Stour in Kent, a few sea trout ascending the latter as far as Chislet. In Sussex the Arun, Western Rother and Ouse are visited by this fish, the Ouse being noted for its runs of big sea trout. It has been suggested that very heavy sea trout captured off the mouth of the Thames and the Kentish coast off Margate may have been on long migration from the Ouse or some other south-coast river that yields big sea trout.

Like sea trout, brown trout are highly regarded as angling fish. Some idea of the extent to which they are sought after can be gained from the fact that, in 1971, 6,985 anglers fished for trout at Weir Wood Reservoir, Forest Row, Sussex, which is only one of a number of south-eastern lakes and reservoirs stocked with these fish. Incidentally, they caught a total of 1,248 brown trout and 4,883 rainbow trout.

Widespread and abundant throughout the region, particularly in upper fast-flowing reaches of rivers and streams, the brown trout eats many kinds of insects, including may-flies, gnats, midges, beetles, moths, ants, bees and wasps. It also devours snails, small fishes, freshwater shrimps, and various other small animals. The stomach of a 2lb trout contained two water-voles, each 4in long!

I have already mentioned the rainbow trout, a species native to north-western America, where it is called steelhead or cut-throat trout. Introduced into Britain in 1884, it is now wide-spread in the South East, where it is restricted mainly to lakes and reservoirs. Rainbows, which are able to withstand a degree of water pollution that defeats brown trout, are known to have bred in the Malling Bourne in Kent, lakes or ponds near Hasle-mere, Surrey, the river Wey in Surrey, a small stream near Chichester, West Sussex, and in a quarry pond in Kent.

The American brook trout (or speckled charr), a native of eastern North America whose food is similar to that of the brown trout, has bred in a chain of spring-fed pools near Haslemere, but attempts to introduce the European huchen, or Danube salmon, to the Thames about 1904 were unsuccessful.

The grayling, with its silvery and grey-green tints and violet stripes, is missing from early twentieth-century lists for the South East. However, Maitland's distribution map for this inhabitant of clear, swift-running, stony waters shows that it is now widely but locally distributed in the region. It is, for example, found in the Medway near Ashurst, in the Wey near Godalming and in the West Sussex Rother. Although the two species will live successfully in certain waters, there are other places where the oft-despised grayling may compete with trout and where the introduction of grayling may be followed by a decline in the trout population. Grayling have been described as being 'extremely unselective' and 'exceptionally unfastidious' in their feeding-habits, their food including insect larvae and nymphs, freshwater shrimps, flies, worms, fish-fry, snails and even common shrews.

Smelt

Like the grayling and so many more species, the smelt affects the populations of certain other fish, its diet including young sprats, herring, whiting and gobies. An inshore fish, it enters estuaries and slightly brackish waters of the lower reaches of rivers to spawn. It used to ascend the Thames as high up as Richmond, and 'John Bickerdyke' stated more than sixty years ago that the smelt was plentiful in the Thames and Medway estuaries. Nowadays this species, which is capable of adapting to life in freshwater, and of giving rise to landlocked populations, is taken in one or two Kentish estuaries. Pollution and overfishing (smelt are delicious to eat) are thought to have reduced the numbers of smelt in the region.

Pike

One of our largest native freshwater fishes, the pike has, not without reason, been called the tyrant of the rivers—the freshwater wolf. Widespread and common in the South East, the pike inhabits rivers, streams, lakes and ponds where, once it has reached a length of 8in, it feeds mainly on fish. Lying concealed amongst weeds, it seizes passing fish, taking advantage of its ability to travel over short distances at nearly 9ft per second and making full use of its large strong teeth. Most of our freshwater fish form part of the pike's diet at some time or other. Over 300 minnows were found inside a pike caught on the Thames.

Apart from fish, the pike takes frogs and newts, ducklings and other small swimming birds, and such small mammals as rats and voles. The effect of pike on the fauna of freshwater can be gauged from the fact that the annual average consumption of a 4½lb specimen is 47lb.

Looking down the list of big pike in the *Woodbine Angling Yearbook 1972*, one finds a number of interesting records from our region, including a 36lb fish from near Maidstone, one of 35lb from a Maidstone lake, and a 33lb specimen from a Kent lake. Until its controversial dismissal from the official list in 1968 the British record pike was one captured in 1945 at a weight of 47lb 11oz. On its removal the pike record was opened for claims at a weight of 41lb.

The carp family

On the deletion of the 1945 pike from the British Record Fish List, a famous 44lb carp (1952) became the biggest coarse fish in the list. Though not reaching this weight, there are large carp in rivers, lakes, pits and ponds in the South East, where this introduced Asiatic species is widespread and common. Since 1954 the lower Thames has yielded a 17½lb carp at Chertsey and a 10lb example at Richmond, while the Canbury Gardens stretch, where warm water comes in from Kingston Power Station, has produced several large carp, including a 17lb 2oz specimen.

Carp may be seen spawning amongst water weeds in shallow water or at the water's edge in May or June, and on sunny summer days quiet observers may watch carp rising to the surface of ponds and pits to bask in the sunshine, and hear them smacking their lips as they take gulps of air. Carp have a varied diet which includes larvae, molluscs, crustaceans, insects and plant matter. In winter carp living in waters that are liable to become frozen cease feeding and bury themselves deep in the mud until the thaw sets in.

The Crucian carp, another introduced species, is not so widespread in the region as the common carp, being recorded most commonly in Sussex. It is able to survive in low oxygen conditions in small ponds, muddy pools and ditches. The Crucian carp does not reach the size of the common carp, but some fine specimens over 4lb in weight have been taken from lakes in Kent and Surrey.

The goldfish and the Crucian carp are very similar and closely related, some experts regarding them as subspecies of the one species *Carassius carassius*. Like its close relative, the goldfish is able to live in low oxygen conditions. Though usually thought of as a domesticated fish of aquaria, garden ponds and lakes, it does live 'wild' in the Thames in artificially warmed water.

There is nothing 'tame' about the wary barbel, a species of clear, swift-flowing, sandy or gravelled rivers, whose strong fighting quality appeals to anglers. The Thames is the great barbel river, but the fish is rare in Kent, though there are some in the upper Medway and in the Stour near Canterbury. In Sussex barbel were introduced into the rivers Ouse and Rother in 1972. A bottom feeder, active mostly at night, the barbel takes larvae of aquatic insects, worms, freshwater shrimps and snails. During April and May barbel migrate upstream in shoals to the spawning beds, gravelly shallows where the yellowish eggs lodge between stones or adhere to their surfaces.

Looking somewhat like a miniature barbel, the gudgeon is widespread and common in the South East. Usually found in shoals, it inhabits sandy and gravelled rivers, ponds and lakes. The main items of diet of this hardy bottom-feeder are insect larvae, freshwater shrimps and snails.

Like the carp, the tench can live in pits, ponds, and small, still waters that may show oxygen deficiency. It also frequents weedy lakes and slow-flowing rivers where mud and silt are able to accumulate. In these deposits this sluggish fish finds food and sanctuary from its enemy the pike. Here, too, it can bury itself, lying dormant in severe winters and during conditions of extreme drought. Adult tench eat bottom-living insect larvae and pond snails, the young feeding on algae, water fleas and water mites. Widespread in the region, this species has been taken in the Cuckmere, the lower reaches of the Beult, the lower Thames, and many landlocked waters.

On Maitland's distribution map for the silver bream (also known as white bream and bream flat) the South-East region is completely blank, but elsewhere there are records of this species from the Thames at Egham and from the Stour in Kent, a county where it is officially stated to be uncommon. Here is an opportunity for angling naturalists to help establish the true status of the silver bream in the region.

Fairly common and widespread here, the common or bronze bream prefers warm waters, quiet lakes and slow-flowing rivers in whose muddy stretches it may gorge upon larvae, worms and pea-mussels. When the water temperature drops bream stop feeding, fasting completely during a really cold winter. On sunny summer days cautious observers may watch these timid, wary fish enjoying the warmth near the surface.

The bleak, dear old Izaak Walton's 'freshwater-sprat' (an apt name if there ever was), is a naturally lively and gregarious little fish of surface waters. Though apparently absent from large parts of the South East, it does occur in all three counties. A species of clear lakes and clean slow-running rivers, it is found in the Arun at Pulborough and in the Thames at such places as Walton and Richmond. The bleak feeds on insects, fish spawn and young fish, and is itself consumed by pike, perch and trout. In eastern Europe this silver-bellied little imp is commercially exploited, its scales being used in the manufacture of artificial pearls.

The minnow is another small species which features in the diet of such fish as chub, eels, perch, trout and, as already mentioned, pike. A common and fairly widespread inhabitant of south-eastern streams, it is also preyed on by kingfishers and other fish-eating birds.

With brownish green, brassy yellow and blood red among its colours, the deep-bodied rudd is not only brilliantly coloured but handsome—the handsomest British freshwater fish in the opinion of many people. A species of rich lakes, ponds, canals and quiet, slow-flowing rivers where weeds abound, the gregarious and sociable rudd is widespread over our three counties, living in the Thames, the Kentish Cray and Royal Military Canal, the larger streams of the Pevensey Levels of Sussex, and many other waters. Adult and larval insects, blood-worms, freshwater shrimps, and plants are all included in its diet, and occasionally large rudd eat young or small fish. The observer may see rudd basking in the sunshine on hot summer days, sometimes with part of the back above the water, and it is often possible to attract them with a few dry bread crusts which, like bleak, they will nibble.

Rudd, inhabitants of middle and surface layers, will frequently share a stretch of water with roach, fish living usually near the bottom. These species take similar food. Widespread and quite common in the South East, the roach prefers quiet waters of ponds, lakes, canals and slow-flowing rivers. Coarse anglers, among whom this species is the most popular fish, have taken some particularly fine roach from the Thames and the Kentish Stour.

Another valued angling fish, the chub occurs in most parts of the region, being most abundant in fast-flowing reaches of rivers. The Adur, Arun, Beult, Ouse, Rother, Thames and Wey are some of the rivers where it is found. In summer it may often be seen floating near the surface of quiet waters, but winter finds it resorting to deeper places where it is safe against cold conditions. Small chub eat larvae, freshwater shrimps, mollusc, algae, plant material, and some fish. When a chub has grown to a length of about 8in, fish form up to about a third of its food, vegetable matter also being important in the diet of larger chub. One observer saw a chub seize a rotting bulrush stem, about 2ft

Page 143 Curlew, in the South East its wild, nostalgic cry is familiar at the coast, but its position as a breeding species is precarious here

Page 144 Shelduck; this striking, goose-like duck is increasing and may be spreading inland as a breeding bird

long, and slowly swallow it. Chub are eaten by herons, king-fishers, otters, pike, trout, and several other predators.

Closely related to the chub, the dace also is important to anglers, and it is yet another fish whose feeding acts as a check on populations of certain insects, molluscs, leeches, algae and a number of other living organisms. Well represented in the region, dace do not neglect the river-bed, but spend much time near the surface of streams and fast-running reaches of rivers, including the Kentish Eden, Mole and Thames. The dace is occasionally still known as the dare or dart, names referring to its graceful dashing movements.

Stone loach

Related to the carp family, the stone (or common) loach is a scaleless fish which is fairly common in stony rivers and streams in many parts of the South East. By day it seeks shelter under large stones or in dense weed-beds, emerging at night or on dull days to search for insect larvae and nymphs and certain other foods. The stone loach is of great interest to naturalists and anglers because, being fairly sensitive to pollution, its absence or presence can be used to indicate the presence or absence respectively of river pollution.

Although it is not included in the south-eastern portion of Maitland's distribution map for the species, the spined loach is said to occur in the Kent River Authority area. It would be interesting to know more about the distribution of this mainly nocturnal, burrowing fish in the region.

Eel

The eel is widespread and common in our three counties, being found in estuaries, rivers, streams, lakes and ponds. Familiarity should never be allowed to breed contempt where this fish is concerned, for its life history is truly spectacular. It starts in the western Atlantic in spring or early summer when spawning

I

takes place. Eventually minute leaf-like larvae, known as lepto-
cephali, appear and soon they begin to drift in north-easterly
moving currents towards our European coasts. After 2½ to 3
years the young fish arrive here, having undergone metamorphosis and become elvers. Easily recognised as tiny eels, many of
them ascend our rivers in spring or summer.

The young eels feed and grow, living largely on insect larvae
and molluscs, and being known during this freshwater stage as
'yellow eels'. After an average of nine (males) or twelve (females)
years 'yellow eels' begin their migration to the sea, sometimes
travelling overland when the ground is sufficiently wet. By the
time they reach river mouths in September or October they have
become 'silver eels', their colour having changed. They now re-
enter the sea and migrate to the spawning grounds where, having spawned, they probably die, leaving their offspring to make
the long journey that they themselves undertook when young.

Sticklebacks

These adaptable and ferocious little fish—tiddlers as they are so
often called—are sometimes included in the diet of large eels.
Two species occur in the region, the three-spined stickleback
appearing to be commoner and much more widespread than the
ten-spined stickleback. Both species are found in estuaries and
many types of freshwater habitats, but the three-spined stickleback also inhabits sea shore pools and is occasionally found well
out to sea.

The male stickleback builds a nest of plant fibres cemented
together with a kidney secretion, the three-spined species selecting a shallow depression on the bottom, the ten-spined species
building some inches above the bottom in dense weeds. The
male in brighter breeding coloration drives the female into the
nest to lay her eggs and then guards it and the young for a time.

The stickleback's food varies according to its habitat, but
small forms of animal life make up the bulk of it. In its turn it
is taken by kingfishers, herons, otters and such fish as perch and

pike, but the effect of its spines has enabled many a stickleback to escape from the mouth of a larger fish.

Bass

Often called 'salmon bass' or 'white salmon', the bass is found off the coast of the region and in estuaries. It sometimes ventures up certain rivers, being taken for example in the Arun between Ford and Arundel Bridge from June to August. Little is known about several aspects of the life history of this species, and it is to be hoped that facts result from the bass-tagging programme being operated by the Ministry of Agriculture, Fisheries and Food Laboratory at Lowestoft. Several hundred bass have already been tagged and released in an effort to establish migratory patterns and growth rates.

Perch and ruffe

Conspicuously coloured and handsome, the perch is a fairly widespread and common inhabitant of south-eastern pits, lakes and rivers. It prefers slow-flowing stretches of rivers, selecting hiding-places in tangles of weeds and roots, darting out to seize small fish and the fry of large species, including its own. Such other food as leeches, freshwater shrimps and larvae are also eaten.

In the spring perch lay long ribbon-like threads of spawn, in which the eggs are embedded at regular intervals. Waterfowl and fish enjoy eating this spawn.

Unlike the perch, whose life-span commonly extends to ten years, the ruffe or pope, a small perch-like fish, usually lives to four years of age. A species of still and slow-flowing waters, it has a somewhat local distribution in the South East and appears to be absent from large areas.

Grey mullets

There are three British species of grey mullets, namely thick-lipped mullet, thin-lipped mullet, and golden mullet. But a glance at Maitland's maps for these fish, which are completely blank for the British Isles, will show that there is great scope for angling naturalists and other observers to help fill in the details of their distribution.

The *Victoria County Histories* include both thin-lipped and thick-lipped mullets for Kent (1908) and Sussex (1905), and a recent official list of Sussex fish species features the thick-lipped mullet, probably the most common and widespread of our British grey mullets.

Grey mullets are, strictly speaking, sea fish and they occur in shallow inshore water along the coast of our region. They venture into estuaries and travel some way up several of our rivers. Grey mullet frequent the Kentish Stour as far as Chislet, the Arun between Ford and Arundel Bridge, and the Adur near Lancing College.

On sunny summer days mullet may often be seen near the surface in places like these, patrolling the river, sometimes at speed, causing a swell or bow wave in front of them, 'arrowing', as it is called. Freshwater anglers frequently speak of mullet as being fiery and unpredictable. Certainly they will leap from the water when a net is closing in upon them. Their food consists mainly of algae and diatoms, but molluscs have been found in large mullet.

Bullhead

Also known as miller's thumb and noggle-head, the bullhead, a flattened, broad-headed little fish, is fairly widespread and common in the region. It prefers stony and gravelled streams and rivers where it can shelter under stones, emerging at night or during dull weather to feed on crustaceans, and insects and their nymphs and larvae. Normally solitary, bullheads pair off in early spring. The female deposits her yellow eggs on the underside of a stone beneath which the male has excavated a small hole. He guards the eggs for about four weeks until they hatch, when the tiny fry scatter to places of safety under stones.

Flounder

An important food fish in parts of Europe, the flounder is a flatfish whose eyed side varies in colour according to the bottom on which it lives. Thus on dark-coloured mud of estuaries it is blackish or dull brown. Elsewhere the eyed side may be greyish or dull green, speckled or marbled. The blind side is usually dead white, but flounders are occasionally coloured on both sides.

The flounder returns to the sea to spawn, but it is essentially an estuarine fish, though it can live in fresh water for a time. It is common off the coast of the region and in estuaries and lower reaches of several south-eastern rivers. In the mid-eighteenth century the Water Bailiff of the City of London boasted of the Thames as an excellent nourisher and a 'speedy breeder' of flounders, and it is said that these fish were found as far upstream as Hampton Court before the construction of Teddington Lock.

The flounder may reach a length of about 20in. A two-pounder is quite a good specimen, but a 4lb 12oz flounder was taken in Pagham Harbour in 1968.

Scientific names of fish mentioned in this chapter:

Barbel, *Barbus barbus*
Bass, *Dicentrarchus labrax*
Bass, salmon. *See* Bass
Bleak, *Alburnus alburnus*
Bream, bronze. *See* Bream, common
Bream, common, *Abramis brama*
Bream, silver, *Blicca bjoerkna*

Bream, white. *See* Bream, silver
Bream, flat. *See* Bream, silver
Bullhead, *Cottus gobio*
Carp, common, *Cyprinus carpio*
Carp, crucian, *Carassius carassius*
Charr, speckled. *See* Trout, American brook
Chub, *Leuciscus cephalus*
Dace, *Leuciscus leuciscus*
Dare. *See* Dace
Dart. *See* Dace
Eel, *Anguilla anguilla*
Flounder, *Platichthys flesus*
Goldfish, *Carassius auratus*
Grayling, *Thymallus thymallus*
Gudgeon, *Gobio gobio*
Huchen, *Hucho hucho*
Lampern. *See* Lamprey, river
Lamprey, brook, *Lampetra planeri*
Lamprey, river, *Lampetra fluviatilis*
Lamprey, sea, *Petromyzon marinus*
Loach, common. *See* Loach, stone
Loach, spined, *Cobitis taenia*
Loach, stone, *Noemacheilus barbatulus*
Miller's thumb. *See* Bullhead
Minnow, *Phoxinus phoxinus*
Mullet, golden, *Chelon auratus*
Mullet, thick-lipped, *Crenimugil labrosus*
Mullet, thin-lipped, *Chelon ramada*
Mullets, grey. *See* above three entries
Noggle-head. *See* Bullhead
Perch, *Perca fluviatilis*
Pike, *Esox lucius*
Pope. *See* Ruffe
Roach, *Rutilus rutilus*
Rudd, *Scardinius erythrophthalmus*
Ruffe, *Gymnocephalus cernua*

Salmon, *Salmo salar*
Salmon, Danube. *See* Huchen
Salmon, white. *See* Bass
Shad, allis, *Alosa alosa*
Shad, twaite, *Alosa fallax*
Smelt, *Osmerus eperlanus*
Stickleback, 3-spined, *Gasterosteus aculeatus*
Stickleback, 10-spined, *Pungitius pungitius*
Tench, *Tinca tinca*
Trout, American brook, *Salvelinus fontinalis*
Trout, brown, *Salmo trutta fario*
Trout, cut-throat. *See* Trout, rainbow
Trout, rainbow, *Salmo gairdneri*
Trout, sea, *Salmo trutta trutta*
Trout, steelhead. *See* Trout, rainbow

CHAPTER SIX

Other wildlife

*Reptiles and amphibians—Snails and slugs—Butterflies
—Moths—Damselflies and dragonflies—Bumblebees and
cuckoo bumblebees—Grasshoppers, crickets and cock-
roachs—Other creatures*

IN RECENT YEARS several of our reptiles and amphibians have
become scarcer. A widespread fear of these animals has often led
people to kill snakes, while the use of herbicidal sprays may well
have had a harmful effect on toads and frogs. Cultivation,
dredging and certain 'tidying-up' operations have modified or
destroyed habitats, rendering them unsuitable for some species.
More frogs and toads are killed crossing roads nowadays and in
Surrey I found another hazard to frogs in the form of a steep-
sided garden pond which they could not leave until a sloping
wooden plank was provided.

Reptiles and amphibians

Slow-worm *Anguis fragilis*: also called blind-worm, this legless
lizard is present in good numbers in many places. Inhabiting
churchyards, heaths, woods and hedgerows, it feeds on worms,
slugs and insects.

Viviparous lizard *Lacerta vivipara*: fairly common and widely
distributed, this hardy species has been seen sunning itself in
sheltered spots where snow has just melted.

Sand lizard *L. agilis*: an inhabitant of heaths and sandy places,
this species has become much scarcer, though it can still be seen

in one small area north of Lewes and near Frensham Little Pond.
Bright green in the breeding season, the male should not be
confused with the introduced green lizard *L. viridis*.

Grass snake *Natrix natrix*: the yellow or orange collar has earned
this abundant and widespread reptile the names of ringed snake
and ring-neck. Seen in hedgerows and open woods in spring, it
is known to migrate annually to and from marshes, where it
feeds on frogs, fish and newts. In the South East the grass snake
usually mates between early April and mid-May. In July the
female lays from twenty to twenty-five eggs in a heap of damp
manure, sawdust or vegetation whose decay provides the
warmth which, like moisture, is essential to incubation. When,
in six to nine weeks, the eggs hatch, the young snake uses the
egg-tooth above its mouth to cut an escape-slit in the leathery
skin of the egg.

Smooth snake *Coronella austriaca*: sometimes mistaken for the
adder, despite the different pattern on its back, the more slender
body and tail, the shape of the head, and its smooth and polished
appearance. Occurs on dry heaths and in open woods in Surrey
and Sussex, particularly along their borders with Hampshire,
but it is rare.

Adder or viper *Vipera berus*: much in evidence in spring, this
venomous snake is still widespread and numerous on heaths,
banks and scrubby downland slopes. It should not be picked up
and handled except by experts.

The amphibians present in the South East are listed below:

Warty or great-crested newt *Triturus cristatus*: widely distributed
in ponds and in marshland drainage dykes (eg Rye area).

Smooth newt *T. vulgaris*: although found under logs and stones
on land for much of the year, it resorts to ponds and ditches
during the mating period. Its numbers may be declining in parts
of the region.

Palmate newt *T. helveticus*: often found in the same places as the warty newt, but its distribution needs further study.

Common toad *Bufo bufo*: widely, though unevenly, distributed.

Natterjack *B. calamita*: also called running toad, it has been recorded from all three counties of the region, but is now very rare in the sandy country of west Surrey.

Common frog *Rana temporaria*: still seen in many places, but numbers appear to have declined steadily, if not rapidly, in recent years. Experienced naturalists have expressed the opinion that, though frogs and toads still spawn, many of the young do not seem to survive to maturity.

Edible frog *R. esculenta*: this active and thoroughly aquatic species is still being introduced to ponds. Recorded from Surrey (eg Ham Common and Richmond Park) and Kent, from where it is spreading into north-east Sussex.

Marsh frog *R. ridibunda*: this native of eastern Europe was originally introduced into Kent in February 1935, when twelve specimens from Hungary were liberated in a garden pond at Stone-in-Oxney on the edge of the Romney Marshes. The district, with its numerous drainage dykes, suited the species and it spread over a wide area, its territory now including a stretch of the Sussex coastal marshes.

Snails and slugs

As is the case with ponds, streams and other watery places, the marshland dykes are very suitable for freshwater snails when there is little or no human interference and when there are good growths of water-plants. Ramshorn snails (*Planorbis* species) and pond snails (*Lymnaea* species) are common in such habitats. Recorded from all three counties, the liver fluke snail *Lymnaea truncatula* is not only the smallest *Lymnaea* but the intermediate host of the liver fluke *Fasciola hepatica*, a parasitic flatworm that is very harmful to sheep.

The hard water of the South East suits certain of the larger freshwater mussels *Unionacea* whose early larval phase is parasitic on the gills or fins of fish. Painter's mussel *Unio pictorum* is known from several places, including the river Wey and the Cut Mill Pond at Puttenham, Surrey. *U. tumidus* inhabits rivers and canals but needs fresher, cleaner water than its much commoner relative. Swan mussel *Anodonta cygnea* is found in rivers and such other stretches of water as the Cut Mill Pond and Ruxley gravel pits near Sidcup.

Other freshwater bivalves represented here are the pea cockles, *Sphaerium* and *Pisidium* species, inhabitants of rivers and canals. *S. rivicola* is the largest *Sphaerium*, while *P. amnicum* is our biggest member of the genus *Pisidium*. The zebra mussel *Dreissena polymorpha*, a brown-and-yellow shell first found in Britain in 1824 in the Surrey Commercial Docks, is a particularly interesting member of the south-eastern fauna, it being the only non-marine mussel to pass through a free-swimming larval stage.

As south-eastern rivers approach the coast one finds semi-marine and salt-marsh snails such as the primitive pulmonate *Phytia myosotis*, the spire shell *Hydrobia ventrosa*, and the dun sentinel or rond snail *Assiminea grayana*, all inhabitants of brackish water. Here, too, lives Jenkins's spire shell *Potamopyrgus jenkinsi*, a species that has spread into fresh running water.

Some of the region's slugs and land snails live in wet places. The hardy slug *Agriolimax laevis* and the snails *Monacha granulata* and *Zonitoides nitidus* represent these hygrophiles, the last-named sometimes being almost amphibious.

Well able to thrive in much drier conditions are certain of the numerous snails of chalkland habitats where calcium carbonate, an important constituent of shells, is abundant. Typical xerophiles are three of the so-called sandhill snails, *Helicella caperata*, *H. itala* and *H. virgata*, thick-shelled creatures of dry grassy places. Living underground in calcareous soil, the blind white snail *Cecilioides acicula* has a thin, transparent shell.

The snails *Clausilia bidentata*, *Cochlicopa lubrica*, *Oxychilus alliarius* (garlic snail), *O. cellarius* and *Vitrina pellucida* are among the species of sward and moss, while the garden snail *Helix aspersa*, the white-lipped banded snail *H. hortensis* and the dark-lipped banded snail *H. nemoralis* occur in rank sward. Much variation is found within populations of both the white-lipped and the dark-lipped banded snails, which also live in woods and other places. They can be banded or unbanded, and the dark bands may be on yellow, brown or pink backgrounds. It has been shown that, in the case of the dark-lipped species, this variation is closely associated with the surroundings in which the snails live. Thus unbanded forms are heavily outnumbered by banded ones in hedgerows, where the interplay of light and shade create patterns that help to conceal the banded varieties from thrushes and other enemies. In deciduous woods the proportion of brown-shelled forms, as compared with yellow-shelled ones, is high in winter, but the balance changes in favour of yellow-shelled forms in spring, when the brown shades of earth and dead leaves give way to the green tints of new leaves and fresh young growth. The large thick-shelled Roman snail *H. pomatia* is found in rank sward and scrub. But, like the round-mouthed snail *Pomatias elegans*, an inhabitant of scrubby and mossy places, it is confined to soils rich in calcium carbonate.

In beechwoods on the chalk one encounters *Ena montana*, with its rich brown shell, on the ground or on the trees themselves, and the smaller, pale brown *E. obscura* among fallen leaves and moss. A very rare species of old downland woods, the cheese snail *Helicodonta obvoluta* has long, curved whitish hairs on its bright brown shell.

Away from the chalkland one usually finds far fewer species. There are, in fact, places on the central Wealden ridge and the Greensand soils where very few slugs and snails are seen. The glossy-shelled *Zonitoides excavatus*, our only lime-avoiding land-snail, and *Hygromia subrufescens*, a small shell of damp old woodland, may be found there.

Characteristic of 'wild' places, *H. subrufescens* is considered to be an anthropophobe, a species intolerant of human interference. *Acanthinula aculeata*, a minute brown woodland snail, and the slugs *Limax tenellus* and *L. cinereoniger*, both local species of old woods, are similarly regarded.

On the other hand, the naturalist who searches large and neglected old gardens may find such anthropophiles as the garden snail *Helix aspersa* (already mentioned as a species of rank sward) and the strawberry snail *Hygromia striolata*. The slugs of such gardens include *Arion hortensis*, a very common pest of crops, the earthworm-eating *Testacella haliotidea* and *T. scutulum*, and two *Milax* species, the black *M. gagates* and the pale brownish-grey *M. sowerbyi*. In one garden, where the weight of live slugs at most seasons of the year was equivalent to two good-sized rabbits, it was calculated that slugs were eating at least 1½cwt of material a year.

Butterflies

Adonis blue *Lysandra bellargus*: confined to places on the chalk downs where its food plant, horseshoe vetch *Hippocrepis comosa*, grows.

Bath white *Pontia daplidice*: a rare immigrant.

Brimstone *Gonepteryx rhamni*: a widespread insect of hedgerows, thickets and woodland rides where either of our buckthorns grow (purging buckthorn *Rhamnus catharticus* and alder buckthorn *Frangula alnus*). After hibernating in an evergreen bush the adult butterfly reappears early in the year, a true harbinger of spring.

Brown argus *Aricia agestis*: absent from much of the region, it occurs where rock-rose *Helianthemum chamaecistus* and stork's bill *Erodium cicutarium* grow, downs, rough grassy slopes and sandy places being among its habitats.

Brown hairstreak *Thecla betulae*: this secretive species of thickets, hedges and woodland margins where blackthorn *Prunus spinosa* grows is most often found in the west of the region.

Camberwell beauty *Nymphalis antiopa*: a rare immigrant.

Chalkhill blue *Lysandra coridon*: like the adonis blue, whose food plant it shares, this charming species belongs to chalk downs.

Clouded yellow *Colias croceus*: a spring immigrant which sometimes breeds here in the summer but does not normally survive the winter.

Comma *Polygonia c-album*: a fairly common and widely distributed inhabitant of woods (its original home), gardens and lanes where winged adults hibernate among dead leaves or exposed on branches.

Common blue *Polyommatus icarus*: generally distributed, being found on downs and hillsides, on the coast and in rough fields and meadows.

Dark green fritillary *Argynnis aglaia*: it seems to have vanished from its coastal haunts in Kent, but may still be seen flying fast and powerfully across the downs.

Dingy skipper *Erynnis tages*: more common in the west than elsewhere in the region, this little dark brown butterfly is seen on hillsides and in meadows, rough fields and open woods.

Duke of Burgundy fritillary *Hamearis lucina*: found in north Kent and towards the west of the region, where it flies along paths and clearings in woods containing its food plants, cowslip *Primula veris* and primrose *P. vulgaris*.

Essex skipper *Thymelicus lineola*: resembling the small skipper, with which it was confused until 1890, it is fairly common along the sea-walls and dykes of the Kent coast and it is also seen in inland localities, particularly in the north of the region.

Gatekeeper—*see* Hedge brown.

Grayling *Eumensis semele*: often seen sunning itself on downs and dry heaths mainly in the north and west of the region.

Green hairstreak *Callophrys rubi*: a species of hedges, scrub, the margins of woods and other places where one finds broom

Sarothamnus scoparius and dyer's greenweed *Genista tinctoria*.

Green-veined white *Pieris napi*: a common and widespread resident whose numbers are sometimes increased by immigrants from the continent.

Grizzled skipper *Pyrgus malvae*: meadows, rough fields, hillsides and open woods are the habitats of this blackish, white-spotted species which is more widespread in the west of the region.

Heath fritillary *Melitaea athalia*: this woodland species, whose main food plant is cow-wheat *Melampyrum pratense*, has suffered from the activities of unscrupulous collectors. In the South East it is practically confined to a small area in north Kent.

Hedge brown *Maniola tithonus*: also called 'gatekeeper' because of its love of sunny openings where gates occur in hedges, this is a widespread bright brown butterfly of lanes, hedgerows, open woods and woodland margins.

High brown fritillary *Argynnis cydippe*: restricted to certain woods where the larvae feed upon violets (*Viola* species) and the adult insects roost in the higher branches on dull days and at night.

Holly blue *Celastrina argiolus*: this widespread insect of woods and garden produces two broods, the first flying in April and May, the other in July and early August.

Large pearl-bordered fritillary *Argynnis euphrosyne*: a tawny-hued species that flies in sunny clearings in certain of the region's larger woods between April and June.

Large skipper *Ochlodes venata*: a widely distributed tawny-yellow butterfly of grassy slopes, meadows, rough fields, woodland paths and sea cliffs.

Large tortoiseshell *Nymphalis polychloros*: a rare species found in woods and along roads where its food plant, elm (*Ulmus* species), occurs.

Large white *Pieris brassicae*: also known as cabbage white, this common resident produces larvae that are destructive to cabbages, cauliflowers and similar vegetables. Its numbers are sometimes reinforced by vast swarms of immigrants.

Long-tailed blue *Lampides boeticus*: a very rare immigrant.

Marbled white *Melanargia galathea*: a local insect of downland, meadows, rough fields and hillsides.

Marsh fritillary *Euphydryas aurinia*: confined largely to the west of the region, it forms colonies in marshes and damp meadows where devil's-bit scabious *Succisa pratensis* grows. It is also occasionally seen on the chalk in places where its food plant is abundant.

Meadow brown *Maniola jurtina*: abundant in many parts of the region in meadows and rough fields, on hillsides and along hedgerows where it finds bramble blossom particularly attractive.

Monarch *Danaus plexippus*: this large, powerful-flying North American species is a very rare immigrant which does not breed here.

Orange-tip *Anthocharis cardamines*: called 'April's butterfly' though it flies also in May and June, this delicate species is common and generally distributed.

Painted lady *Vanessa cardui*: a regular spring immigrant whose numbers vary from year to year, it breeds here in summer along hedgerows and in other places where thistles flourish, but does not survive our winter.

Pale clouded yellow *Colias hyale*: a spring immigrant reaching Britain less often than the clouded yellow.

Peacock *Nymphalis io*: a common and widespread resident whose larvae feed on stinging nettles (*Urtica*). Like the small tortoiseshell, it is attracted to such garden flowers as Japanese stonecrop, Michaelmas daisy and Buddleia.

Purple emperor *Apatura iris*: a beautiful high-flying species of large oakwoods, especially of those in the west of the region. Its larvae feed on sallow (goat willow, *Salix*), the winged insect itself being attracted to sap exuded by wounded trees and to juices of dead rabbits and other corpses.

Purple hairstreak *Thecla quercus*: feeding on oak (*Quercus*), it is seen in and around large oakwoods, most modern records being from the western part of the region.

Queen of Spain fritillary *Argynnis lathonia*: a very rare immigrant.

Red admiral *Vanessa atalanta*: this common spring immigrant breeds here in the summer on nettles (*Urtica*). A few survive our winter, but the majority die or return to the south in the autumn.

Ringlet *Aphantopus hyperanthus*: grassy lanes and meadows and woodland rides are the habitats of this fairly common species whose larvae feed on grasses.

Short-tailed blue *Everes argiades*: a very rare casual immigrant.

Silver-spotted skipper *Hesperia comma*: a speciality of the downland chalk, this species has disappeared from some of its old haunts, but is reported to have recovered on the North Downs in recent years.

Silver-studded blue *Plebejus argus*: the dry sandy heaths of Surrey are a stronghold of this species whose larvae are 'farmed' by ants, which milk the larval honey-gland at intervals.

Silver-washed fritillary *Argynnis paphia*: a woodland species which, like the less widespread high brown fritillary, roosts in trees.

Small blue *Cupido minimus*: lives in well-defined colonies on downs and rough grassy slopes on the Kent coast and also in the west of the region.

K

Small copper *Lycaena phlaeas*: widely distributed over downs, hillsides, rough fields, meadows and coast.

Small heath *Coenonympha pamphilus*: often regarded as the most numerous of all British butterflies, it is seen on rough ground, marshes, hillsides and the seashore.

Small pearl-bordered fritillary *Argynnis selene*: an inhabitant of damp woods that also flies in open marshes.

Small skipper *Thymelicus sylvestris*: widespread, frequenting such varied habitats as inland forest rides and sea-walls and dykes along the coast.

Small tortoiseshell *Aglais urticae*: a common and widespread resident whose larvae feed on stinging nettles (*Urtica*). Like the peacock butterfly, it has been caught in lightships off the South East coast.

Small white *Pieris rapae*: a common and widespread resident of cultivated areas whose numbers are augmented by immigrants.

Speckled wood *Pararge aegeria*: frequent in shady places in woods and along hedgerows and wooded lanes.

Swallowtail *Papilio machaon*: swallowtails occasionally recorded in the South East as 'casuals' or 'migrants' appear to belong to a distinct continental race, the British swallowtail being confined to the Norfolk fens.

Wall *Pararge megera*: well distributed in sunny lanes, hedgerows and field margins.

White admiral *Limenitis camilla*: despite its decline in recent years, this woodland butterfly, the emblem of the Sussex Naturalists' Trust, still occurs in several parts of the South East.

White-letter hairstreak *Strymonidia w-album*: present in a number of woods and lanes where elms (*Ulmus* species) grow, it should be sought at privet and bramble blossom.

Wood white *Leptidea sinapis*: a weak-flying inhabitant of woods whose recent 'recovery' has been reported from the Weald.

Moths

Generations of naturalists have collected and studied south-eastern moths, but there is still great scope for moth-hunters, especially for those who are prepared to tackle the smaller species.

The following moths are typical of the Chalk, though some of them are also found elsewhere:

Barred sallow *Tiliacea aurago*, chalk carpet *Ortholitha bipunctaria*, dusky sallow *Eremobia ochroleuca*, five-spot burnet *Zygaena trifolii*, galium carpet *Epirrhoë galiata*, light-feathered rustic *Agrotis cinerea*, reddish light arches *Apamea sublustris*, six-spot burnet *Zygaena filipendulae*, and treble-bar *Anaitis plagiata*.

Species likely to be found on heaths and heathy commons include: beautiful yellow underwing *Anarta myrtilli*, common heath *Ematurga atomaria*, emperor *Saturnia pavonia*, heath rustic *Amathes agathina*, July highflyer *Hydriomena furcata*, ling pug *Eupithecia goossensiata*, map-winged swift *Hepialus fusconebulosa*, and true-lover's knot *Lycophotia porphyrea*.

The following may also occur in such places when pines are present: barred red *Ellopia fasciaria*, bordered white *Bupalus piniaria*, grey pine carpet *Thera obeliscata*, pine beauty *Panolis flammea* and pine carpet *Thera firmata*.

R. E. Scott, who operated moth traps there in recent years, found many interesting species on the shingle area at Dungeness. According to this observer, the commonest hawk moth at Dungeness was the small elephant hawk *Deilephila porcellus*. This was followed by the eyed hawk *Smerinthus ocellata* and poplar hawk *Laothoe populi*. The pebble prominent *Notodonta ziczac* was found to be by far the most numerous of the prominents.

Dungeness Foreland has, of course, long been regarded as an area of great entomological importance. The grass eggar moth *Lasiocampa trifolii* ab *flava* occurs there in an ochreous yellow form which is believed to be found nowhere else in the world.

Several other moths either occur only at Dungeness and a very few other places, or are found there in distinct forms.

South-eastern woods also support many interesting moths. One Kent wood, now a nature reserve, is the home of two aspen-feeding species, the Clifden nonpareil *Catocala fraxini* and the lesser belle *Colobochyla salicalis*. Here, too, one finds the scarce merveille-du-jour *Moma alpium*, the dark crimson underwing *Catocala sponsa*, the light crimson underwing *C. promissa*, the lunar doublestripe *Minucia lunaris* and the sub-angled wave *Scopula nigropunctata*.

Another south-eastern nature reserve is the sole site in Britain of the Lewes wave *Scopula immorata*, a moth first recorded from the area in 1887. Here scrub is being controlled to maintain the essential grass-heath conditions, an activity which has already proved particularly beneficial to the insects.

Damselflies and dragonflies

Most of the forty-five British species of the *Odonata* have been recorded from the region. South-eastern damselflies include the comparatively large demoiselle agrion *Agrion virgo*, the banded agrion *A. splendens*, the green lestes *Lestes sponsa*, the large red damselfly *Pyrrhosoma nymphula*, the common ischnura *Ischnura elegans* (which may be confused with the scarce ischnura *I. pumilio*), and the common blue damselfly *Enallagma cyathigerum*, which is often found far from water.

Hawker dragonflies are represented here by the emperor dragonfly *Anax imperator*, the largest British hawker dragonfly, the hairy dragonfly *Brachytron pratense*, a real spring insect, and several others, including four species of *Aeshna*, namely *cyanea*, *juncea*, *grandis* and *mixta*.

Twelve of the darter dragonflies have been recorded from the South East. Perhaps the best known are the four-spotted libellula *Libellula quadrimaculata*, the largest of the family, the broad-bodied libellula *L. depressa*, the common sympetrum *Sympetrum striolatum*, the 'fire-bellies' of Sussex, and the ruddy sympetrum

S. sanguineum. Recently the migrant yellow-winged sympetrum *S. flaveolum* surprised Surrey naturalists by returning after some twenty years to Thursley, an area where much thought has been given to problems of managing wet areas so as to avoid unchecked plant succession, drying out and loss of habitats.

Bumblebees and cuckoo bumblebees

Loss of habitats, such as hedgerows, banks and pieces of rough ground, is probably among the causes of the decline in the bumblebee population which is commonly supposed to have taken place in recent years. Fortunately there are corners of the region where these efficient pollinating insects may still be seen.

The seven British species regarded as 'generally common' occur here. Of these *Bombus terrestris*, whose queen has a buff-coloured tail in Britain, prefers underground nest-sites, as does the red-tailed *B. lapidarius*. *B. hortorum* also nests underground but favours sites with much shorter approach tunnels. *B. lucorum*, another underground nesting species, is a prolific pollen-storing species. At Englefield Green in Surrey an entomologist watched a red-tailed bumblebee queen using her jaws and legs to excavate an underground tunnel and saw her disappear from sight in just over half an hour. *B. agrorum* and *B. ruderarius* usually nest on the surface of the ground, often under moss or grass, while *B. pratorum* uses such varied habitats as disused birds' nests and old mattresses.

The distribution of a particular bumblebee is sometimes followed very closely by the cuckoo bumblebee which is parasitic on it. Resembling bumblebees of the genus *Bombus*, these parasites belong to the genus *Psithyrus*. Each of the six British cuckoo bumblebees has been recorded from the South East, though *P. bohemicus* is very rare here despite the abundance of its host, *B. lucorum*.

Grasshoppers, crickets and cockroaches

Some thirty-four species of *Orthoptera* have been recorded from the South East, about nineteen of them from all of the five vice-counties used for scientific recording.

Three of the nine British species of cockroaches have been widely reported here. They are the common cockroach *Blatta orientalis* (the 'blackbeetle' of dirty places indoors), the German cockroach *Blattella germanica*, and the tawny cockroach *Ectobius pallidus*, a species of low vegetation in woods and on downs, commons and heaths. Crickets are still fairly well represented in the region and it is encouraging to note that the Sussex Naturalists' Trust has been able to make a gentleman's agreement over a small area of land which supports a flourishing colony of the rare, flightless field cricket *Gryllus campestris*.

Bush-crickets, or 'long-horned grasshoppers', with long, thread-like antennae, occur here. The oak bush-cricket *Meconema thalassinum* lives in oak and other trees, but thick vegetation is the habitat of the great green bush-cricket *Tettigonia viridissima*. The dark bush-cricket *Pholidoptera griseoaptera* hides in nettle-beds and hedgerows, while the bog bush-cricket *Metrioptera brachyptera* favours heath bogs and damp vegetation. Somewhat neglected gardens are among the haunts of the speckled bush-cricket *Leptophyes punctatissima*.

Other species that are widespread in the region are: short-winged cone-head *Conocephalus dorsalis*, mole-cricket *Gryllotalpa gryllotalpa*, stripe-winged grasshopper *Stenobothrus lineatus*, common green grasshopper *Omocestus viridulus*, woodland grasshopper *O. rufipes*, common field grasshopper *Chorthippus brunneus*, meadow grasshopper *C. parallelus*, lesser marsh grasshopper *C. albomarginatus*, rufous grasshopper *Gomphocerippus rufus*, mottled grasshopper *Myrmeleotettix maculatus*, common ground-hopper *Tetrix undulata*, slender ground-hopper *T. subulata*.

Other creatures

That the south-eastern fauna has many other claims on the naturalist is evident from reports in both serious and popular natural history publications. One can only hope that more and more naturalists will take up the less popular groups.

The scope for spider studies is evident when one reads that well over one hundred spiders of five different species were found in a Surrey telephone kiosk where they were feeding on insects attracted inside by the light.

The discovery in 1968 of a new British centipede *Brachyschendyla dentata* in soil samples collected in the centres of Guildford and Haslemere emphasises the value of detailed soil sampling, while the capture in a bus travelling from Maidstone to Sevenoaks of an introduced Australian beetle *Paratillus carus* shows the need for vigilance on the part of the naturalist.

CHAPTER SEVEN

Places to visit

*Nature reserves and other sites—Museums—The South
Downs Way—The North Downs Way—Maps*

NATURALISTS RESIDENT IN the region will already have their
favourite places for nature rambles. It is hoped, however, that
these alphabetical lists of open spaces and nature reserves will
assist not only visitors to the South East but residents venturing
into less familiar areas.

The following abbreviations are used:

ccos	County Council Open Space (access limited in a few cases to footpaths and bridleways)
FC	Forestry Commission
LNR	Local Nature Reserve
m	mile(s)
NCSE	Nature Conservancy South-East Region
NNR	National Nature Reserve
NT	National Trust
PAR	Permit required to visit parts away from the rights of way
PVR	Permit to visit required
RSPB	Royal Society for the Protection of Birds
SNT	Sussex Naturalists' Trust

The map references in the list enable places to be located on
maps of the Ordnance Survey, the AA *Road Book*, and others
bearing the National Grid.

Readers are reminded that no attempt should be made to
collect specimens unless written permission has previously been

obtained. Where necessary, applications for permits should be made *well in advance* of the dates of proposed visits.

Nature reserves and other sites

Kent

In addition to those listed, the Kent Trust for Nature Conservation has about 30 reserves in the county. Details of these are issued to members of the Trust.

Bedgebury National Pinetum, TQ 7233. 5m NW of Hawkhurst. FC. Conifers, forest plots, forest trail nearby. Guide available (see Bibliography).

Blean Woods, TR 1664–1864. 3m NW of Canterbury. NNR, PAR refer NCSE. 164 acres, deciduous woodland, interesting insects.

Challock Forest. FC.
  (a) TR 020503. King's Wood forest trail. Varied conifers and sweet chestnut coppice.
  (b) TR 140440. Lyminge forest trail.

Dungeness, TR 0916. PVR, refer RSPB Reserves Department. 1,233 acres, part of the unique shingle peninsula jutting out into the English Channel. Also two fresh water lakes. Shore and water birds. Famous migration point.

Ham Street Woods, TR 0033–0034. 5m S of Ashford. NNR, PAR refer NCSE. 240 acres, 237 acres woodland, characteristic insect fauna and flora of coppice-with-standards type of woodland.

High Halstow, TQ 7876. 6m NE of Rochester. NNR (Nature Reserve agreement with RSPB), PAR refer NCSE. 130 acres, 117 acres woodland, large heronry at Northward Hill. Birds: breeding species, migrants, winter visitors.

Hothfield Common, TQ 9745. 3m N of Ashford. LNR. 143 acres, heath and bog, including fine valley bogs.

Mildmay Forest, TQ 495725. 1m S of Bexley station. FC. Forest trail through Joydens Wood, varied trees, wild plants.

Northward Hill, TQ 7876. PVR refer RSPB Reserves Department.

Part of the High Halstow NNR (see above). Overlooks N Kent marshes.

Orlestone Forest, TQ 983349. FC. Forest trail through mixed woodlands on Weald Clay.

Parson's Marsh, TQ 471528. 1½m S of Brasted. NT, 18½ acres woodland (oak).

Petts Wood, TQ 450687. Between Chislehurst and Orpington on W side of A208. NT, 88 acres wood (oak) and heath. Adjoining and to W of Petts Wood is Hawkwood (also NT).

Scord's Wood, TQ 477520. 2m S of Brasted, W of Ide Hill. NT, 69 acres woodland (sessile oak).

South Swale, TR 035648. On S bank of the Swale, by Graveney Marshes. LNR. Keep to public footpath along S boundary. 1,500 acres. Waders and wildfowl.

Stodmarsh, TR 2261. 5m NE of Canterbury. NNR. Public access route, PAR, NCSE. 402 acres, shallow pools, reed beds, swamps, damp meadows. Wildfowl and other wetland birds.

Swanscombe Skull Site, TQ 6174. 4m E of Dartford. NNR. Inner enclosure by permit, refer NCSE. 5 acres, site of human skull dating from at least 100,000 years ago.

Toys Hill, TQ 465517. 2½m S of Brasted. NT. 93 acres woodland (beech) and heathland.

Wye and Crundale Downs, TR 0745. 5m NE of Ashford. NNR, PAR refer NCSE. 250 acres, including chalk downland, scrub and 100 acres of woodland. Nature trail.

Surrey

Abinger Roughs, TQ 103479. Just N of Abinger Hammer. NT, 98 acres, wooded ridge.

Albury Heath, TQ 0647. 90 acres.

Ashtead Common, TQ 1759. 458 acres.

Ashtead Park, TQ 1959. 45 acres.

Bagshot Heath, SU 8963. 186 acres.

Banstead Downs, TQ 2561. 430 acres.

Banstead Heath and Burgh Heath, TQ 2457. 847 acres.

Barfold Copse, Haslemere, SU 9232. PVR, refer RSPB Reserves Dept. 13 acres, mixed woodland, many typical birds.

Bisley Common, Reidon Hill area, SU 9558. CCOS, 66 acres.

Bisley Green and Bisley Common, SU 9559. CCOS, 51 acres.

Blackheath, TQ 036466. ½m S of Chilworth & Albury station. Part NT, 408 acres, heather and pines.

Blindley Heath, TQ 3745. 72 acres.

Bookham and Banks Commons, TQ 1256. 2½m W of Leatherhead. NT, 447 acres, wooded common. For reports on this area see publications of London Natural History Society.

Box Hill, TQ 1751. NT, 841 acres, down and woodland. Juniper Hall, at foot of Little Switzerland Valley, is Field Centre of the Council for the Promotion of Field Studies. Mickleham Downs, 73 acres, adjoins Box Hill property.

Broadstreet, Backside and Rydeshill Commons, SU 955510. CCOS, 414 acres.

Broadwater, SU 9845. 66 acres.

Burners Heath, SU 948550. CCOS, 14 acres.

Chertsey Mead, TQ 0666. 160 acres.

Chinthurst Hill, TQ 013455. CCOS, 38 acres.

Chobham Common, SU 9765. Part CCOS, 1,629 acres.

Churt Common, SU 8640, 202 acres.

Clandon Downs, TQ 0651. 60 acres.

Claremont Woods, TQ 130632. On S edge of Esher. NT, 49 acres, woods, lake.

Clasford Common, SU 950525. By A323. CCOS, 10 acres.

Coldharbour Common, TQ 1544. 42 acres.

Cranleigh Common, TQ 053394. 62 acres.

Crooksbury Common, SU 8945. 160 acres.

Crooksbury Hill, SU 8846. CCOS, 45 acres.

Ditton Common, TQ 1566. 79 acres.

Dunsfold Common, TQ 0136. 102 acres.

East Sheen Common, TQ 197746. Adjoining Richmond Park on the N. NT, 53 acres.

Epsom Common, TQ 1961. 430 acres.

Epsom Downs and Walton Downs, TQ 2258. 607 acres.

Esher and Arbrook Commons, TQ 1462. 367 acres.

Fairmile Common, TQ 1262. 137 acres.

Farley Heath, TQ 055455. 140 acres.

Farnham Park, SU 8448. 308 acres.

Frensham Common, SU 8540. Astride the A287. NT, 665 acres, heathland, includes most of Frensham Great Pond.

Frith Wood and Puplet Wood, TQ 3761. 117 acres.

Froggitts Heath and Newchapel Green, TQ 3642. 51 acres.

Gatton, TQ 265522. 1½m NE of Reigate. NT, 212½ acres, on N slopes of North Downs.

Hackhurst Down, TQ 092486. ½m NE of Gomshall & Shere station. Part NT, CCOS and other land open to the public. 223 acres, on North Downs.

Hankley Common, SU 8842. 745 acres.

Harewoods, TQ 3347. At Outwood, 3m SE of Redhill. NT, public access to Outwood Common and to parts of woodlands.

Headley Heath, TQ 2053. 4m S of Epsom. NT, 482 acres.

Hill Park Estate, Tatsfield, TQ 425559. CCOS, 60 acres.

Hill Top, Chaldon, TQ 314535. CCOS, 40 acres.

Hindhead—NT

    Grayswood Common, SU 917344. 1m N of Haslemere. NT, 16½ acres, but 48 acres also open.

    Hindhead, Inval & Weydown Commons &c, SU 8935. 12m SW of Guildford on both sides of the A3. 1,076 acres, connected commons, heath and wood.

    Nutcombe Down, Tyndall Wood & Craig's Wood, SU 887352. 95 acres, heath and wooded valley.

    Polecat Copse, SU 885338. 36 acres, mainly woodland.

    Stoatley Green, SU 896339. 1m N of Haslemere station. 5 acres.

    Woodcock Bottom, SU 880361. ¼m NW of Hindhead. 105 acres, wood and heath.

Hog's Back (2 areas), SU 8048 and SU 9248. CCOS, 32 acres.

Holmwood Common, TQ 1746. 1m S of Dorking on both sides of the A24. NT, 630 acres, common land, much wooded.

Horsell Common, SU 96/TQ 06. 730 acres.

Hurtwood and Winterfold Heath, TQ 0944. 1,850 acres.

Hydon's Ball and Hydon Heath, TQ 978396. 3m S of Godalming. NT, 125 acres, heath and woodland.

Kew (ex Surrey): Royal Botanic Gardens, TQ 1876. National collection of living and dried plants, botanical museums.

Laleham Park, TQ 053680. 70 acres.

Leith Hill—NT

   Duke's Warren, TQ 142442. 4m SW of Dorking. 193 acres, heath and wood.

   Leith Hill Place, TQ 134424. 421 acres, woods.

   Leith Hill Summit, TQ 139432. 1m SW of Coldharbour. 5 acres on highest point in SE England (965ft).

   Mosses Wood, TQ 146432. 69 acres, bluebell woods.

   Severell's Copse, TQ 130454. At Friday Street, 1m E of Abinger. 59 acres, partly wooded, borders Friday Street lake.

Limpsfield Commons, TQ 4252. 496 acres.

Littlefield Common, SU 9653. CCOS, 42 acres.

Little Heath, SU 975628. CCOS, 12 acres.

Littleworth Common, TQ 1565. 121 acres.

Lucas Green and West End Common, SU 9450. CCOS, 70 acres.

Merrow Common, TQ 0352. 50 acres.

Merrow Downs, TQ 04/05. 318 acres.

Milford Green and Coxhill Green, SU 985610. CCOS, 51 acres.

Netley Park, TQ 078484. At E end of Shere. NT, woodlands. Netley Plantation, CCOS, 23 acres, adjoins this.

Newlands Corner and Silent Pool, TQ 0449. Access area managed by County Council, 260 acres.

Nonsuch Park, TQ 2364. 255 acres.

Norbury Park, TQ 1654. CCOS, 1,251 acres.

Ockham and Wisley Commons and Chatley Heath, TQ 0851. CCOS, 728 acres.

Oxshott Heath, TQ 1461. 260 acres.

Park Downs, Banstead Wood and Fames Rough, TQ 2757. 433 acres.

Petridge Wood Common, TQ 278472. 52 acres.

Polesden Lacey, TQ 140533. 2½m S of Bookham station. On N

slope of North Downs. NT, large gardens, beech walks.

Pray Heath, SU 9555. 41 acres.

Puttenham Common, SU 915465. Access area managed by County Council, 470 acres.

Puttenham Heath, SU 945475. 62 acres.

Pyrford Common, TQ 0359. 50 acres.

Ranmore Common, TQ 1451. 2m NW of Dorking. NT, 472 acres, wooded common. Denbies Hillside, 245 acres on S slopes of North Downs, adjoins the Common.

Ranmore section of Abinger Forest, TQ 126501. FC. Forest trail, varied woodlands on crest of North Downs.

Redhill and Earlswood Commons, TQ 2749. 334 acres.

Reigate Heath, TQ 235503. 128 acres.

Reigate and Colley Hills, TQ 250520. NT, 149½ acres, open down, copse and beechwood on summit and escarpment of North Downs, with views to Leith Hill and South Downs.

Reigate Park and Priory, TQ 245495. 139 acres.

Richmond Park (ex Surrey), TQ 2073. Deer, birds.

Rickford Common, SU 9754. CCOS, 52 acres.

Ripley Green, TQ 055570. CCOS, 68 acres.

Royal and Bagmoor Commons, SU 9243. 200 acres.

Run Common, TQ 0342. 51 acres.

Runnymede, TQ 007720. On the Thames, ½m above Runnymede Bridge. NT, 188 acres, historic meadows. Cooper's Hill Slopes (also NT), 110 acres, overlooks the meadows.

Rushett Common, TQ 0242. 43 acres.

St John's Lye, SU 9857. 61 acres.

St Martha, TQ 025482. Access area managed by County Council, 93 acres.

Sandhills Common, SU 942381. ½m W of Witley station. NT, 11½ acres.

Selhurst Common, etc, TQ 0241. 58 acres.

Selsdon Wood, TQ 363615. 3m SE of Croydon. NT, 198½ acres.

Shabden Park Estate, TQ 2756. CCOS, 447 acres.

Shalford Park, SU 995485. 142 acres.

Sheepleas Estate, TQ 085515. CCOS, 269 acres.

Sheets Heath, SU 9557. 64 acres.
Shepperton Manor, TQ 080667. CCOS, 23 acres.
Shere Heath, TQ 0747. 50 acres.
Shortwood Common, TQ 0572. 60 acres.
Smarts Heath, SU 9856. 63 acres.
Smithwood Common, TQ 0541. 113 acres.
Staines Moor, TQ 0373. 290 acres.
Stanners Hill, TQ 0063. CCOS, 104 acres.
Stoke Park, TQ 0151. 176 acres.
Stony Jump, SU 870396. 1m E of Churt (A287). NT, 37 acres, heather hill, path to summit.
Streets Heath, SU 950614. CCOS, 19 acres.
Stringers Common, SU 9953. CCOS, 73 acres.
Swan Barn Farm, SU 912328. On E edge of Haslemere. NT, farmland, woodland and chestnut coppice, access by footpath.
Tandridge: Hanging Wood, TQ 368536. NT, 4½ acres, on S slope of North Downs.
The Chantries, TQ 0148. 459 acres.
The Nower and Milton Heath, TQ 1648. 60 acres.
Thursley and Ockley Commons, SU 9141. 893 acres.
Tilburstow Common, TQ 355504. 63 acres. Adjoins Tilburstow Hill. CCOS, 23 acres.
Walton Heath and Little Heath, TQ 2354. 468 acres.
Warren Farm, TQ 1953. CCOS, 117 acres.
West End Common, TQ 125635. 129 acres.
Westfield Common, TQ 0056. 71 acres.
West Hanger, Combe Bottom, TQ 065485. CCOS, 89 acres.
Weybridge: The Heath, TQ 0864. 58 acres.
Whitmoor Common, SU 995535. CCOS, 432 acres.
Windsor Great Park, SU 96/97. 850 acres.
Winkworth Arboretum, SU 990412. 3m SE of Godalming. NT, 99 acres, rare trees and shrubs, lake, grass field.
Wisley: Royal Horticultural Society's gardens and arboretum. TQ 0658.
Witley and Milford Commons, SU 9240. ½m SW of Milford. Part NT, 774 acres.

Woldingham: South Hawke, TQ 373540. 1½m S of Woldingham. NT, 4½ acres, down and woodland on a ridge of the North Downs.

Wonersh and Shalford Commons, TQ 0246. 137 acres.

Wotton and Abinger Commons, TQ 1345. 721 acres.

Sussex

Amberley Wild Brooks, TQ 031148. SNT, 10½ acres. Alluvial plain on E side of the Arun, with aquatic flora and fauna.

Balcombe Marsh, TQ 315284. SNT, PVR, ½ acre of marsh.

Beachy Head, TV 5995. Downland and cliffs.

Black Down (Sussex and Surrey), SU 9230. 1m SE of Haslemere. NT, 602 acres, including the highest point in Sussex (918ft).

Cissbury Ring, TQ 140082. 3m N of Worthing. NT, 80 acres, chalk grassland, site of early Neolithic flint-mining industry.

Crowlink, Michel Dene and Went Hill, TV 5497. 5m W of Eastbourne. NT, 632 acres, cliff, down and farmland. Adjoins Birling Gap (also NT), TV 555957. 66½ acres, ¼m of chalk cliffs.

Duncton Chalk Pit, SU 961163. SNT, 5 acres, chalk woodland and scrub surrounding a disused chalk quarry.

Durford Heath, etc. SU 788256. 2½m NE of Petersfield, on S side of A3. NT, 64 acres, including 58 acres of heath.

Fairlight, TQ 884127. 4½m E of Hastings (A259). NT, 228 acres, including 58 acres of cliffland.

Flatropers Wood, TQ 862234. SNT, PVR. 86 acres, mainly oak/birch woodland.

Friston Forest—FC

(a) *Main block*, TV 526996. Forest trail through plantations of pine and beech on South Downs.

(b) *Wilmington block*, TQ 558078. Forest trail, 3m long.

Highdown Hill, TQ 092043. 3m NW of Worthing. NT, 50 acres, important archaeological site.

Kingley Vale, SU 8210. 4m NW of Chichester. NNR, PAR, refer NCSE. 351 acres, including 'the finest natural yew forest in Europe', open downland turf and chalk heath.

Lavington Common, SU 950190. 1m W of Petworth. NT, 77 acres, heather and pines.

Lullington Heath, TQ 5302. 4m NE of Seaford. NNR, PAR, refer NCSE. 155 acres, chalk heath vegetation.

Marley (Sussex and Surrey)—NT

  Kingsley Green Common, SU 894306. 2m S of Haslemere. 7 acres.

  Marley Common and Wood, SU 887313. 132 acres, high common and steep woodland.

  Marley Coombe, SU 887317. 19 acres.

  Marley Heights, SU 893300. Viewpoint.

  Shottermill Ponds, SU 883324. Two hammer ponds and adjacent land.

Nap Wood, TQ 582330. 4m S of Tunbridge Wells on A267. NT, leased to SNT. Unrestricted access on Sundays, April–October. PVR at other times. 110 acres of central wealden oakwood.

Newbury Pond, Cuckfield, TQ 306243. SNT, 1 acre, pond and surrounding land.

Newtimber Hill, TQ 2712. 5m NW of Brighton. NT, 238 acres, down and woodland, views of the Weald and the sea.

North Common, Chailey, TQ 3818. 6½m NNW of Lewes. LNR, 429 acres, dry and wet heathland, small patches of sphagnum bog.

Nymans Gardens, TQ 265294. 4½m S of Crawley. NT, rare conifers, shrubs and plants. Open April–October.

Pagham Harbour, around SZ 8796. 5m W of Bognor Regis. LNR, tidal mud, reed beds, pools and ditches. Birds include winter visitors and passage migrants.

Pagham Lagoon, SZ 884970. Adjoins Pagham Harbour LNR. SNT, winter visitors include duck, grebes, divers, waders and rails.

Rye Harbour, TQ 91. LNR, 209 acres. Terns, shingle plants.

Saddlescombe Chalk Pit, TQ 267122. On W side of Newtimber Hill. SNT. Contains juniper scrub.

St Leonard's Forest, near Horsham, TQ 141255. FC. Forest trail, 2m long, in Marlpost Wood.

L

St Leonard's Forest. SNT has Nature Reserve agreement with FC over:

  (a) *Sheepwash Ghyll*, TQ 207298. Central wealden ghyll woodland on Tunbridge Wells sandstone.
  (b) *Mick's Cross*, TQ 214304. 12 acres, plateau woodland, beech with holly.
  (c) *Lily Beds*, TQ 212308. Colony of lily-of-the-valley *Convallaria majalis*.

Seaford Head, TV 49. LNR, 76 acres, chalk downland, narrow valley with mixed scrub, cliffs and shingle beach.

Selsfield Common, TQ 348345. 4m SW of East Grinstead. NT, 7 acres.

Selwyns Wood, TQ 552207. SNT. Access: May–October, Wed 2–6 pm, Sat & Sun 10 am–6 pm. 28 acres, chestnut coppice, mixed deciduous woodland, conifers.

Sheffield Park Gardens, TQ 415240. ½m from Sheffield Park station (Bluebell line). NT, gardens and five lakes, rhododendrons and azaleas in late spring, rare trees, splendid autumn colours. Open April–November (April: Wed, Sat, Sun only).

Shoreham Gap, TQ 2209. 2m NE of Shoreham. NT, 596 acres of down.

Slindon Estate, SU 9608. 6m N of Bognor Regis on A29. NT, 3,503 acres, farm and woodland. The park containing the Slindon beeches is open daily (no access for vehicles). Access to remainder of estate by public footpath only.

Slindon Forest—FC, young beechwoods and conifers on slopes of South Downs, N and NW of Slindon village.

  (a) *Northwood Forest Trail*, starts at SU 938106.
  (b) *Marden Forest Trail*, starts at car park, 1m N of Stoughton, at SU 814126.
  (c) *Selhurst Park Forest Trail*, SU 925128.

Sullington Warren, TQ 096144. ½m E of Storrington. NT, 28 acres, with wide views of the North and South Downs.

The Mens, TQ 028242. SNT, 116 acres, mixed oak and beech woodland, which is said to approach the ancient Wealden forest in character more closely than any other known woodland.

The Warren, Chailey, TQ 405216. SNT, PVR. 6½ acres, dry and damp heathland.
Vert Wood, TQ 511149. SNT, PVR. 5 acres, birds, butterflies, moths.
Vinehall Forest, TQ 764203. FC. Forest trail, pinewoods, larchwoods, sweet chestnut coppices.
Wakehurst Place, TQ 3331. 1½m NW of Ardingly on B2028. NT, 120 acres of gardens containing an important collection of exotic trees, shrubs and other plants. Open daily (except Christmas Day).
Warren Hill, TQ 116139. 8m N of Worthing. NT, 247 acres, including Washington Common. Views of South Downs.
Welch's Common, SU 982175. SNT. A mixture of dry heathland and wet mixed alder and sallow carr.
West Wittering: East Head, SZ 766990. On E side of Chichester harbour entrance. NT, 76 acres, 1¼m coastline, dunes, saltings, sandy beaches.
Woods Mill, Henfield, TQ 218137. SNT headquarters and education centre in 15-acre nature reserve (woodland, pasture, marsh, lake, several streams). Nature trail. Open April–September: Tues, Wed, Thurs and Sat 2–6 pm, Sundays and public holidays 11 am–6 pm. Educational visits by appointment.
Woolbeding, SU 8724. 2m NW of Midhurst. NT, 1,035 acres. Public access to the commons (400 acres) and to part of the woodlands.

Museums (with natural history collections):

Kent

Dartford	Dartford Borough Museum, Market Street
Dover	Dover Corporation Museum, Ladywell
Downe	Darwin Museum, Down House
Folkestone	Folkestone Museum and Art Gallery, Grace Hill
Maidstone	Maidstone Museum and Art Gallery, St Faith's Street
Tunbridge Wells	Museum and Art Gallery, Mount Pleasant

Surrey
Camberley	Camberley Museum, Municipal Buildings
Haslemere	Haslemere Educational Museum, High Street

Sussex
Bexhill	Bexhill Museum, Egerton Park
Bognor Regis	Museum Collection of Natural History, Lyon Street
Brighton	Booth Museum of British Birds, Dyke Road
	Museum and Art Gallery, North Gate House, Church Street
Hastings	Public Museum and Art Gallery, John's Place, Cambridge Road

The South Downs Way

This long-distance bridleway, which is open to walkers, riders and cyclists, was established under the provisions of the National Parks and Access to the Countryside Act of 1949.

It runs for some 80 miles between Eastbourne and the Hampshire border, mostly along the ridge of the South Downs. Special oak signs and small stone plinths bearing the words 'South Downs Way' are used to mark the route. These signs incorporate the Countryside Commission's Long-Distance Path waymarking symbol, a stylised acorn, which is also used by itself at intermediate points on gateposts, stiles, walls or fences.

The following publications will also assist the rambler:

Ramblers' Association booklet *The South Downs Way*, by Ernest G. Green, price 5p (plus postage).
Countryside Commission leaflet (with map) *The South Downs Way*. Information sheet on accommodation (with intermediate mileages), which includes addresses of establishments offering stabling or grazing, from the Countryside Commission or from Mr Charles Shippam, Priory Cottage, Boxgrove, Chichester, Sussex (*enclose stamped addressed envelope*).
Sheets 181 (Chichester), 182 (Brighton), and 183 (Eastbourne)

of the Ordnance Survey One-Inch to One-Mile maps (*use the latest edition*).

Detailed maps of the route may be inspected at the offices of the Countryside Commission, the East Sussex County Council and the West Sussex County Council.

The North Downs Way

This scenic route of 141 miles, from Farnham in Surrey to Dover in Kent, follows as far as possible the crest of the North Downs, and in a few places coincides with the medieval Pilgrims' Way from Winchester to Canterbury.

A descriptive leaflet (with map showing the positions of youth hostels) may be obtained from the Countryside Commission.

Maps

The region is covered by the following sheets of the Ordnance Survey one-inch map of Great Britain: 169–73, 181–4. Sheet 17 (South-East England) of the OS quarter-inch series is also useful.

CHAPTER EIGHT

Index of South-Eastern birds

With notes on distribution

SPECIES ARE PLACED in alphabetical order for ease of reference. The number following a name refers to that given in the *Check-List of the Birds of Great Britain and Ireland.** Extreme rarities are omitted. To conserve space, the following abbreviations are used:

GP	Gravel pit(s)	PM	Passage migrant
RES	Reservoir(s)	SF	Sewage farm(s)
SV	Summer visitor	WV	Winter visitor

AUK, LITTLE *Plautus alle*, 226. Irregular WV, sometimes blown inland.

AVOCET *Recurvirostra avosetta*, 185. Scarce, chiefly spring and autumn, mainly coast and vicinity. Bred S Kent/Sussex border until c 1843, pair bred N Kent, 1958.

BEE-EATER *Merops apiaster*, 259. Very rare, mainly spring. 2 pairs raised young, Sussex, 1955.

BITTERN *Botaurus stellaris*, 38. Rare, breeds Stour valley, Kent, where slowly increasing. Visits wet and swampy places, chiefly winter.

    LITTLE *Ixobrychus minutus*, 37. Very rare, summer, coast and vicinity, GP, inland SF.

BLACKBIRD *Turdus merula*, 308. Abundant resident, WV, PM.

BLACKCAP *Sylvia atricapilla*, 343. Abundant SV, few winter here.

*British Ornithologists' Union, London, 1952.

CORNCRAKE *Crex crex*, 125. Few seen April/May, August/ September, mainly on coast.

CRAKE, SPOTTED *Porzana porzana*, 121. Skulking, easily overlooked. Few records, mainly February, April, November, December, coast, inland SF.

CROSSBILL *Loxia curvirostra*, 404. Breeding in scattered localities where conifer seeds are available (also feeds on Whitebeam berries, Surrey). Numbers increased by irregular irruptions across North Sea (eg autumn, 1966).

CROW, CARRION *Corvus corone corone*, 280. Abundant and widespread resident.

   HOODED *Corvus corone cornix*, 281. Uncommon WV, coast, inland SF. Single birds occasionally in summer.

CUCKOO *Cuculus canorus*, 237. Widespread SV.

CURLEW *Numenius arquata*, 150. A few breed, Surrey and Sussex. Some non-breeders present in summer. Many in winter on Swale, Medway, Thames, at Pett Level, Rye and Pagham harbours.

   STONE *Burhinus oedicnemus*, 189. Scarce SV to Dungeness and Sussex Downs, where few pairs nest.

DABCHICK. See under GREBE, LITTLE.

DIVER, BLACK-THROATED *Gavia arctica*, 1. Few seen off coast or at Pagham Harbour in winter. More seen off Dungeness, Selsey Bill, Beachy Head, during passage, spring, autumn.

   GREAT NORTHERN *Gavia immer*, 2. Few visit Pagham and Chichester harbours in winter. Seen off coast on passage.

   RED-THROATED *Gavia stellata*, 4. Small numbers along coast in winter, also seen on passage.

DOTTEREL *Eudromias morinellus*, 142. Scarce, irregular PM, coast, Sussex Downs, spring, autumn.

DOVE, COLLARED *Streptopelia decaocto*, 236. A widespread and abundant resident, which continues to spread. Assembles in winter flocks.

   STOCK *Columba oenas*, 232. A common resident. Winter flocks.

TURTLE *Streptopelia turtur*, 235. Abundant SV. Flocks seen after nesting is completed. Few over winter, Kent, Sussex.

DUCK, LONG-TAILED *Clangula hyemalis*, 61. Scarce PM and WV to coast and vicinity, Surrey RES.

MANDARIN *Aix galericulata*. An imported species, probably escaped from collection at Cobham, Surrey, breeds freely on and near middle reaches of R Mole.

TUFTED *Aythya fuligula*, 56. Resident and common WV to harbours, GP, RES (including Surrey), lakes. Breeds throughout region in limited numbers.

DUNLIN *Calidris alpina*, 178. Common on coast all year round, most numerous autumn/winter. Also visits RES, GP, SF (including Surrey). Does not breed here.

DUNNOCK *Prunella modularis*, 371. Abundant and widespread resident.

EIDER *Somateria mollissima*, 67. WV, coastal waters, few non-breeders remain during summer.

FIELDFARE *Turdus pilaris*, 302. WV, numerous.

FIRECREST *Regulus ignicapillus*, 365. Scarce autumn/winter visitor. Also seen on spring and autumn passage, chiefly at coast.

FLYCATCHER, PIED *Ficedula hypoleuca*, 368. PM, seen spring and autumn, mainly at coast.

SPOTTED *Muscicapa striata*, 366. SV, abundant. Nesting places include parks and gardens.

FULMAR *Fulmarus glacialis*, 26. Formerly largely an offshore bird. Recently gained foothold as breeding species, cliffs at Folkestone/Ramsgate.

GADWALL *Anas strepera*, 49. Small flocks in autumn/winter, lakes, RES, GP, marshes. Few pairs breed, Kent, Surrey. Hand-reared birds released at Sevenoaks.

GANNET *Sula bassana*, 27. Seen offshore in all months, sometimes fishing close offshore.

GARGANEY *Anas querquedula*, 47. SV, breeding Kent, Sussex (Cliffe Peninsula, Stodmarsh, Medway estuary, etc). Passage birds seen at GP, SF.

GODWIT, BAR-TAILED *Limosa lapponica*, 155. WV, abundant, coast

(Swale, Medway estuary, Chichester and Pagham harbours). PM, few non-breeding birds remain in summer.

BLACK-TAILED *Limosa limosa*, 154. Plentiful WV to coast and PM. Pairs present, Kent, in summer, but no definite proof of breeding yet.

GOLDCREST *Regulus regulus*, 364. Fairly abundant resident, breeding Downs, large gardens, woods (fond of conifers). Also seen as PM and WV.

GOLDENEYE *Bucephala clangula*, 60. PM, WV, small groups at coast, estuaries, RES, GP (including Surrey).

GOLDFINCH *Carduelis carduelis*, 393. Abundant resident and PM. Some large autumn/winter flocks.

GOOSANDER *Mergus merganser*, 70. Scarce WV to coastal areas and inland waters, numbers increase in severe weather.

GOOSE, BARNACLE *Branta leucopsis*, 81. Occasional WV. Some 'escapes' seen in the region.

BEAN *Anser fabalis*, 78. Uncommon and irregular WV.

BRENT *Branta bernicla*, 80. PM, WV (Thames, Swale, Medway, Chichester and Pagham harbours).

CANADA *Branta canadensis*, 82. This introduced species continues to increase, breeding on lakes and GP, and moving freely about the region.

GREY LAG *Anser anser*, 75. Scarce WV. Introduced free-winged birds are increasing.

PINK-FOOTED *Anser brachyrhynchus*, 78. Uncommon and irregular WV.

WHITE-FRONTED *Anser albifrons*, 76. WV in large and small skeins. Some large counts on Thames, Swale and Medway, January/March.

GREBE, BLACK-NECKED *Podiceps nigricollis*, 8. PM and scarce WV to coast, GP, RES (including Surrey). A few present in summer. One pair bred in Kent, 1964, 1966.

GREAT CRESTED *Podiceps cristatus*, 5. Widespread resident, breeding on all suitable large sheets of water. Winter concentrations on Channel, estuaries, GP, RES.

LITTLE *Podiceps ruficollis*, 9. Well distributed resident, breeding on ponds, GP, rivers, lakes. Autumn/winter concentrations mainly in coastal localities.

RED-NECKED *Podiceps griseigena*, 6. Uncommon PM and WV. Seen on Channel, tidal waters, Surrey RES.

SLAVONIAN *Podiceps auritus*, 7. PM and scarce WV to coastal areas, occasionally seen at inland RES.

GREENFINCH *Chloris chloris*, 392. Abundant resident. Large flocks out of breeding season, many in coastal areas.

GREENSHANK *Tringa nebularia*, 165. PM and irregular WV, coast, inland, GP, SF, RES.

GUILLEMOT *Uria aalge*, 227. Observed at sea most months, many oiled birds seen. Formerly bred on chalk cliffs, Kent and Sussex.

GULL, BLACK-HEADED *Larus ridibundus*, 208. Seen throughout the year, being particularly abundant in winter (coast, towns and suburbs). Breeding colonies at Dungeness and Rye Harbour GP.

COMMON *Larus canus*, 201. Common as WV and PM (coast, inland RES and SF). Breeds (or attempts to) at Dungeness reserve, Lydd ranges, Denge Marsh.

GLAUCOUS *Larus hyperboreus*, 202. Scarce PM and WV, coast or immediate vicinity.

GREAT BLACK-BACKED *Larus marinus*, 198. Common autumn/ spring, less so in summer, at coast and inland (rubbish tips, RES, GP).

HERRING *Larus argentatus*, 200. Resident, nesting on Sussex cliffs and roofs of houses at coast. Also common and widespread WV, coast, inland (rubbish dumps).

ICELAND *Larus glaucoides*, 203. Rare as summer or winter visitor and PM, mainly on coast.

LESSER BLACK-BACKED *Larus fuscus*, 199. Breeds on cliffs, Folkestone and Hastings areas. In autumn/winter large numbers are seen on coast and inland (RES, rubbish dumps, playing fields).

LITTLE *Larus minutus*, 207. Scarce PM and WV, nearly all on coast.

MEDITERRANEAN *Larus melanocephalus*, 205. Rare PM and occasional summer and winter visitor, coast or near.

HARRIER, HEN *Circus cyaneus*, 100. A regular and not uncommon autumn/spring visitor to coastal areas and, less commonly, inland localities.

MARSH *Circus aeruginosus*, 99. In small numbers as PM or WV. Has bred in Kent.

MONTAGU'S *Circus pygargus*, 102. Scarce PM and regular spring and summer visitor. Has bred in small numbers. Coastal levels, marshes, Downland.

HAWFINCH *Coccothraustes coccothraustes*, 391. A shy and retiring resident of well-wooded localities. Forms flocks out of the breeding season.

HERON *Ardea cinerea*, 30. A resident, breeding in some 24 sites. More widespread in autumn/winter when it visits estuaries and the shore. Also seen as PM and WV.

HOBBY *Falco subbuteo*, 104. A few pairs of this summer visitor breed on heathland and in old secluded woods each year.

HOOPOE *Upupa epops*, 261. Seen in very small numbers, March/November, on coast and inland (including Surrey gardens). Has bred.

JACKDAW *Corvus monedula*, 283. Abundant resident, PM and WV.

JAY *Garrulus glandarius*, 286. Abundant resident in woodland areas.

KESTREL *Falco tinnunculus*, 110. A widespread and fairly common resident. Also occurs as PM and WV.

KINGFISHER *Alcedo atthis*, 258. A widespread and not uncommon resident along rivers, streams, lakes. Often visits coast, even beach, August/March.

KITE *Milvus milvus*, 95. A very rare visitor.

KITTIWAKE *Rissa tridactyla*, 211. Present on or off the coast throughout the year. Single birds occasionally seen on Surrey, SF, RES.

KNOT *Calidris canutus*, 169. PM and numerous WV on or near coast. Few seen inland.

LAPWING *Vanellus vanellus*, 133. Nests widely on marshes, grass, arable, SF, GP, RES, on coast and inland. From about June to February occurs in flocks. Some large passage and weather movements occur.

LARK, SHORE *Eremophila alpestris*, 273. Small numbers on coast, September/April.

SKY *Alauda arvensis*, 272. Abundant as resident (breeding in open country), PM and WV.

WOOD *Lullula arborea*, 271. A very local and uncommon breeding bird. Small parties on coast, autumn/winter.

LINNET *Carduelis cannabina*, 395. Present throughout the year, nesting widely on commons, rough bushy ground, low vegetation. Also occurs on passage and as WV.

MAGPIE *Pica pica*, 284. Numerous resident, even visiting and nesting in large gardens.

MALLARD *Anas platyrhynchos*, 45. Abundant and widely distributed resident whose numbers are swollen by influx of autumn/winter visitors. Coastal marshes, estuaries, inland waters.

MARTIN, HOUSE *Delichon urbica*, 276. Numerous SV, nesting on buildings in suburbs, towns and villages and on chalk cliffs.

SAND *Riparia riparia*, 277. Another abundant SV, breeding in sand, gravel and chalk pits, holes and pipes in masonry.

MERGANSER, RED-BREASTED *Mergus serrator*, 69. PM and WV to estuaries. Scarce and occasional on inland waters.

MERLIN *Falco columbarius*, 107. Scarce PM and WV, seen mainly on coast.

MOORHEN *Gallinula chloropus*, 126. A common and widespread resident.

NIGHTINGALE *Luscinia megarhyncha*, 322. SV, breeding fairly commonly in bushes/undergrowth on commons and in open woodland.

NIGHTJAR *Caprimulgus europaeus*, 252. SV to Sussex Downs and interior of region, where it breeds on heaths, commons and in open woodlands.

NUTCRACKER *Nucifraga caryocatactes*, 285. Vagrant from northern conifer forests. An unprecedented irruption in autumn,

1968, brought several to the region (Slender-billed or Siberian form).

NUTHATCH *Sitta europaea*, 296. A common resident in suitably wooded localities. Will breed in garden nest boxes.

ORIOLE, GOLDEN *Oriolus oriolus*, 278. In recent years a few 'singles' seen, May/June, coast and inland. Bred in the region this century.

OSPREY *Pandion haliaetus*, 103. Very scarce PM to coastal and inland waters.

OUZEL, RING *Turdus torquatus*, 307. PM in small numbers. Coast, Downs, few inland.

OWL, BARN *Tyto alba*, 241. A widespread, though no longer plentiful, resident of farmland.

LITTLE *Athene noctua*, 246. A fairly common and widely distributed resident.

LONG-EARED *Asio otus*, 248. A scarce and thinly distributed resident, nesting in wooded areas (prefers conifers).

SHORT-EARED *Asio flammeus*, 249. A few pairs breed in Medway estuary/Sheppey area, where birds are seen throughout the year. Also occurs in small numbers as PM and WV in coastal areas and inland (marshes, heaths, SF, Sussex Downs).

TAWNY *Strix aluco*, 247. Resident in well-wooded areas (with old trees), in places abundant.

OYSTERCATCHER *Haematopus ostralogus*, 131. Nests on shore, harbour islets and fields adjoining coast, where some non-breeders also 'summer'. Abundant PM and WV to estuaries and coastal marshes.

PARTRIDGE, COMMON *Perdix perdix*, 116. A common resident, widely distributed over farm-, marsh- and downland.

RED-LEGGED *Alectoris rufa*, 115. A generally distributed resident, less numerous than the common variety.

PEREGRINE *Falco peregrinus*, 105. Bred until quite recently. 'Singles' now seen, chiefly at coast, usually August/March.

PHALAROPE, GREY *Phalaropus fulicarius*, 187. Rare PM, singly, coast, September/November.

RED-NECKED *Phalaropus lobatus*, 188. Rare PM, singly, coast, May, August/September.

PHEASANT *Phasianus colchicus*, 118. Common and widespread resident.

PIGEON, WOOD *Columba palumbus*, 234. Very common as resident and WV.

PINTAIL *Anas acuta*, 52. Few pairs in breeding season (has bred). Larger numbers seen on passage and as WV to coastal marshes and tidal waters. Occasional inland (RES, SF).

PIPIT, MEADOW *Anthus pratensis*, 373. A somewhat local breeding species of open places. Numerous as PM.

ROCK *Anthus spinoletta petrosus*, 379a. Very few pairs breed, Kent and Sussex (sea cliffs). More numerous as PM and WV, mainly at coast.

TAWNY *Anthus campestris*, 375. Very rare PM, coast or near, autumn.

TREE *Anthus trivialis*, 376. PM and SV, breeding in woodland areas and on commons.

WATER *Anthus spinoletta spinoletta*, 379b. 'Singles' or small numbers at coast and inland (RES, SF), October/April.

PLOVER, GOLDEN *Pluvialis apricaria*, 140. PM and abundant WV to arable land, pasture and downland. Often associates with lapwings.

GREY *Pluvialis squatarola*, 139. PM and common WV to coast. Few non-breeders remain in summer.

KENTISH *Charadrius alexandrinus*, 136. Formerly bred, Kent and Sussex coasts. Now seen in ones and twos, March/May, August/September.

LITTLE RINGED *Charadrius dubius*, 135. PM and SV, breeding in small numbers at inland GP and on other stony ground.

RINGED *Charadrius hiaticula*, 134. Resident, nesting on coast and adjacent pits and diggings. PM and WV. Small numbers seen at inland RES, SF, GP.

POCHARD *Aythya ferina*, 57. Breeds on suitable waters. Seen, sometimes in largish numbers, as WV on coastal and inland waters, including RES, GP.

RED-CRESTED *Netta rufina*, 54. Rare visitor to RES, GP.

PUFFIN *Fratercula arctica*, 230. Scarce and irregular visitor to coastal waters. Sometimes blown inland or washed ashore.

QUAIL *Coturnix coturnix*, 117. Scarce SV to farmland, where it is heard calling from growing corn. Has bred recently.

RAIL, LAND. See under CORNCRAKE.

WATER *Rallus aquaticus*, 120. A scarce breeding species, PM and WV to damp places, including RES, SF, GP, garden pools, village ponds.

RAZORBILL *Alca torda*, 224. PM and WV to coastal waters. Some 'oiled' birds washed ashore.

REDPOLL, LESSER *Carduelis flammea cabaret*, 397. Not uncommon, breeding on heaths and scrubby commons and in forestry plantations. Some largish winter flocks.

REDSHANK *Tringa totanus*, 161. A fairly common resident, breeding on levels, marshes, pastures (some inland), and spending autumn/spring on coast, where influx of winter visitors occurs.

SPOTTED *Tringa erythropus*, 162. PM and WV recorded in every month. Seen, singly or in small flocks, on coast and at inland GP, SF, RES.

REDSTART *Phoenicurus phoenicurus*, 320. SV, nesting where there are old trees. Common as PM on coast.

BLACK *Phoenicurus obscurus*, 321. SV, several pairs breeding, including those at Croydon and Dungeness power stations. Few winter on coast.

REDWING *Turdus iliacus*, 304. Abundant PM and WV. Seen in towns and gardens in really severe weather.

ROBIN *Erithacus rubecula*, 325. Common and widespread resident. Also abundant as PM and WV.

ROOK *Corvus frugilegus*, 282. Common resident, PM and WV.

RUFF *Philomachus pugnex*, 184. WV to coast (some largish flocks) and inland levels, RES, SF. Also PM and occasional SV.

SANDERLING *Crocethia alba*, 181. Abundant PM and WV, mainly on or near coast.

SANDPIPER, COMMON *Tringa hypoleucos*, 159. Recorded in every month, chiefly near coast. Most numerous on autumn passage. Few present in summer and not many wintering birds recorded.

CURLEW *Calidris testacea*, 179. PM, most numerous in autumn, mainly coast and near.

GREEN *Tringa ochropus*, 156. Recorded in every month, mostly near coast. Most common in autumn.

PECTORAL *Calidris melanotas*, 176. Very rare autumn vagrant, coast.

PURPLE *Calidris maritima*, 170. PM and WV in small numbers, coast.

WOOD *Tringa glareola*, 157. Scarce PM, seen near coast and at inland RES, SF, GP.

SCAUP *Aythya marila*, 55. WV and PM. Prefers open sea and waters near the coast, though a few visit inland GP, RES.

SCOTER, COMMON *Melanitta nigra*, 64. WV and PM in fluctuating numbers. Some non-breeders remain in summer. Mainly maritime, but a few seen at inland RES.

VELVET *Melanitta fusca*, 62. PM and WV, wholly maritime.

SHAG *Phalacrocorax aristotelis*, 29. A maritime species recorded in small numbers in most months.

SHEARWATER, BALEARIC *Puffinus puffinus mauretanicus*, 16. Few seen off coast, August/October.

CORY'S *Puffinus diomedea*, 20. Few seen off coast, April, September/November.

MANX *Puffinus puffinus*, 16. Few seen off coast, April/June, August/November.

SOOTY *Puffinus grisea*, 21. Few seen off coast, July/October.

SHELDUCK *Tadorna tadorna*, 73. Breeds commonly on or near parts of coast, but is spreading inland as breeding species. Out of breeding season seen on coastal marshes and mudflats, few at inland RES, GP, SF.

SHOVELER *Anas clypeata*, 53. SV, breeding in several localities. Makes good numbers as PM and WV to coastal marshes, estuaries, and inland RES, GP.

M

SHRIKE, GREAT GREY *Lanius excubitor*, 384. A scarce PM and WV, coast and inland.

RED-BACKED *Lanius cristatus*, 388. Very scarce PM and SV (few nest), coast and inland.

SISKIN *Carduelis spinus*, 394. WV in fluctuating numbers, haunts places with alders and birches, visits large gardens. Some large flocks, but 10–30 usual.

SKUA, ARCTIC *Stercorarius parasiticus*, 193. Seen off coast, April/June, July/November.

GREAT *Stercorarius skua*, 194. Seen off coast, usually small numbers, mainly April/May, July/November.

POMARINE *Stercorarius pomarinus*, 195. Seen off coast on passage, April/May, August/November.

SMEW *Mergus albellus*, 71. Scarce WV, coast and inland RES, GP.

SNIPE *Gallinago gallinago*, 145. Breeds (numbers falling) on marshy ground, rough pastures and commons, levels. Numerous PM and WV, coast and inland (including SF).

JACK *Lymnocryptes minimus*, 147. PM and WV, 'singles' or small numbers, coast and inland (including SF, GP).

SPARROW, HEDGE. See under DUNNOCK.

HOUSE *Passer domesticus*, 424. Very common resident, usually near buildings, but flocks seen in open spaces, late summer/autumn.

TREE *Passer montanus*, 425. Common resident, PM and WV. Some largish autumn/winter flocks.

SPARROWHAWK *Accipiter nisus*, 93. Scarce PM and resident of well-wooded districts, breeding in small numbers.

SPOONBILL *Platalea leucorodia*, 42. Irregular PM, 'singles' on coast, May/June, August/November.

STARLING *Sturnus vulgaris*, 389. Very common resident, PM and WV. Some large flocks seen.

STINT, LITTLE *Calidris minuta*, 171. Scarce PM, most numerous in autumn, chiefly coast, few at inland GP, RES, SF. 'Singles' seen in winter.

TEMMINCK'S *Calidris temminckii*, 173. Very scarce spring and autumn PM, coast.

STONECHAT *Saxicola torquata*, 317. Local resident, breeding on downland, heaths and commons. More widespread as PM and WV.

STORK, WHITE *Ciconia ciconia*, 40. Vagrant to Kent coast, Surrey, April/July.

SWALLOW *Hirundo rustica*, 274. Abundant PM and widespread SV, breeding quite commonly.

SWAN, BEWICK's *Cygnus columbianus*, 86. WV to coast and inland GP, RES, SF, marshes. Usually small numbers.

MUTE *Cygnus olor*, 84. Resident, breeding on most suitable waters. After breeding some move to coast, where non-breeders are seen all year. Immigrants present during severe weather.

WHOOPER *Cygnus cygnus*, 85. Irregular WV, small numbers, coast, inland RES.

SWIFT *Apus apus*, 255. Abundant PM and SV, nesting in buildings.

TEAL *Anas crecca*, 46. Resident, breeding in small numbers, marshes, levels, RES, GP. PM and abundant WV to estuaries, harbours, inland RES.

TERN, ARCTIC *Sterna paradisea*, 218. PM, seen off coast, rarely inland.

BLACK *Chlidonias niger*, 212. PM, 'singles' or small parties, coast and inland GP, RES. Has bred.

COMMON *Sterna hirundo*, 217. Abundant PM and SV. Main breeding colonies at Dungeness, Medway estuary, Cliffe Pools (Kent), and Rye Harbour GP and Chichester Harbour (Sussex).

GULL-BILLED *Gelochelidon nilotica*, 215. 'Singles' off coast, spring and autumn.

LITTLE *Sterna albifrons*, 222. PM and SV. Main breeding colonies at Dungeness, Isle of Sheppey (Kent), and Chichester, Pagham and Rye harbours (Sussex).

ROSEATE *Sterna dougallii*, 219. Few seen, usually on or off coast, April/August.

SANDWICH *Sterna sandvicensis*, 223. Common PM, most on

coast, some at inland GP, RES. Some 'summer' offshore, June/July.

WHITE-WINGED BLACK *Chlidonias leucopterus*, 213. Vagrant, coast, 'singles', August/October in recent years.

THRUSH, MISTLE *Turdus viscivorus*, 301. Widespread resident, PM and WV. After breeding season forms flocks (to c 50) which are seen feeding on berries, even in towns.

SONG *Turdus philomelos*, 303. Widespread resident, PM and WV.

TIT, BEARDED *Panurus biarmicus*, 295. Breeds in reed-beds, Kent (main stronghold in Stour valley), but occasionally seen in other parts of Kent and Sussex.

BLUE *Parus caeruleus*, 289. Common resident, PM and WV.

COAL *Parus ater*, 290. Resident, most abundant among conifers. Visits gardens.

GREAT *Parus major*, 288. An abundant resident.

LONG-TAILED *Aegithalos caudatus*, 294. Resident, common and widespread again, breeding woods, commons, large gardens.

MARSH *Parus palustris*, 292. Resident in wooded areas, rather local.

WILLOW *Parus montanus*, 293. A rather local resident, breeding in damp woods.

TREECREEPER *Certhia familiaris*, 298. Widespread resident of wooded areas.

TURNSTONE *Arenaria interpres*, 143. PM, non-breeding sv (few), and abundant WV, mainly at coast, but few at inland RES, SF.

TWITE *Carduelis flavirostris*, 396. PM and WV, some large flocks on coast, occasional ones and twos inland.

WAGTAIL, BLUE-HEADED *Motacilla flava flava*, 382b. PM in small numbers, mainly on coast.

GREY *Motacilla cinerea*, 381. Numerous and widely distributed, nesting in holes and crevices, on ledges and against banks at outfalls of ponds or lakes, bridges and weirs. Widespread in winter at SF, GP, RES, other inland waters.

PIED *Motacilla alba yarrellii*, 380a. Widespread and abundant resident and PM. Some large roosts and flocks at SF and reed beds, August/April.

WHEATEAR *Oenanthe oenanthe*, 311. PM and SV, a few pairs breeding.

WHIMBREL *Numenius phaeopus*, 151. PM and scarce non-breeding SV, coast and inland, where 'singles' are seen at SF and RES and parties are observed in flight.

WHINCHAT *Saxicola rubetra*, 318. PM seen on coast and inland (SF, RES, Downs), and local summer visitor, few pairs breeding on rough grassland, heaths.

WHITETHROAT *Sylvia communis*, 347. PM and SV, numbers 'crashed' in 1968–9.

LESSER *Sylvia curruca*, 348. PM and widely, but thinly, distributed SV.

WIGEON *Anas penelope*, 50. PM and numerous WV to coast, estuaries, flooded areas, inland RES. A few non-breeders present in summer.

WOODCOCK *Scolopax rusticola*, 148. Widely distributed and fairly numerous resident, breeding in woodlands. PM and WV.

WOODPECKER, GREAT-SPOTTED *Dendrocopus major*, 263. Widespread resident, breeding in wooded districts. Visits gardens.

GREEN *Picus viridis*, 262. Widespread resident, again breeding commonly in wooded areas. Visits large gardens.

LESSER-SPOTTED *Dendrocopus minor*, 264. A widespread resident.

WREN *Troglodytes troglodytes*, 299. An abundant and widespread resident.

WRYNECK *Jynx torquilla*, 265. Scarce PM and declining SV, very few pairs breeding now. Visits Surrey gardens.

YELLOWHAMMER. See under BUNTING, YELLOW.

Bibliography

Alford, D. V. 'Bumble bee distribution maps scheme: progress report for 1970', *Bee World*, 52 (1971), 55–6

Barrington, C. A. *Forestry in the Weald* (Forestry Commission Booklet No 22, 1968)

Blackmore, M. *The Nature Conservancy Handbook* (1968)

British Ecological Society. *Biological Flora of the British Isles* (Accounts of species reprinted from *The Journal of Ecology*, 1941–71)

Clapham, A. R., Tutin, T. G. and Warburg, E. F. *Flora of the British Isles* (Cambridge, 1962)

Conder, P. 'The Wheatear', *Birds*, 2 no 12 (1969), 291–3

Corbet, G. B. 'Provisional distribution maps of British mammals', *Mammal Review*, 1 no 4/5 (1971), 95–142

Council for Nature. Monthly Press Bulletins (1960–5); *Habitat* (1965–71)
  Predatory mammals in Britain (1967)

Countryside Commission. *The North Downs Way* (1970)
  The South Downs Way (1970)

Crowcroft, P. *The life of the shrew* (1957)

des Forges, G. and Harber, D. D. *A guide to the birds of Sussex* (1963)

Dungeness Bird Observatory Committee. *Dungeness Bird Observatory* (1969)

Forestry Commission. *See your forests* (1970)

Free, J. B. and Butler, C. G. *Bumblebees* (1959)

Frewer, G. 'Walking the South Downs', *Birds*, 2 no 10 (1969), 247–50 (79 species of birds seen)

Gallois, R. W. *British Regional Geology, The Wealden District* (1965)

Gillham, E. H. and Homes, R. C. *The birds of the North Kent marshes* (1950)

Gilmour, J. *Wild flowers of the chalk* (1947)

Gooders, J. *Where to watch birds* (1967)

Green, E. G. *The South Downs Way* (Ramblers' Association booklet)

Harrison, J. 'A gravel pit wildfowl reserve', *Birds*, 1 no 3 (1966), 48–51

'Invasion by waxwings', *Birds*, 1 no 5 (1966), 90–2

A wealth of wildfowl (1967)

'Oil pollution fiasco on the Medway estuary', *Birds*, 1 no 7 (1967), 134–6

'The wildfowler as a conservationist', *Birds*, 2 no 8 (1969), 203–5

Harrison, J. M. *The Birds of Kent* (2 vols, 1953)

Heath, J. *Provisional atlas of the insects of the British Isles, Part 1, Butterflies* (Monks Wood Experimental Station, Abbots Ripton, Huntingdon, 1970)

Hudson, W. H. *Nature in Downland* (1900)

Jessup, R. *South East England* ('Ancient peoples and places' series, 1970)

Kent Ornithological Society. *The Kent Bird Report* (1952–70)

Kent Trust for Nature Conservation. Annual Reports (1958–70), *The first ten years, 1958–1968*

London Natural History Society. Annual reports in the *London Naturalist* for the area within twenty miles of St Paul's

The birds of the London area (1964)

Longfield, C. *The dragonflies of the British Isles* (1949)

Lousley, J. E. *Wild flowers of the chalk and limestone* (1950)

Matthews, L. H. *British mammals* (1952)

McMillan, N. F. *British shells* (1968)

Mitchell, A. F. *Short guide to Bedgebury Pinetum and Forest Plots* (1969)

National Trust. *Properties of the National Trust* (1969)

Nature Conservancy. *Evidence of the Nature Conservancy for the public inquiry into the proposed nuclear power station at Dungeness, Kent* (1958)

The management of chalk and limestone grassland for wildlife conservation (1971)

Nature Conservancy (South-east region). *Kingley Vale National Nature Reserve, W. Sussex* (Wye, 1970)

Lullington Heath National Nature Reserve (Wye, 1970)

Newton, I. 'The bullfinch problem', *Birds*, 1 no 4 (1966), 74–7

Olney, P. 'Berries and birds', *Birds*, 1 no 5 (1966), 98–9

Paton, J. A. *Census catalogue of British hepatics* (= liverworts) (1965)

Perring, F. H. and Walters, S. M. (Editors). *Atlas of the British Flora* (1962)

Perring, F. H. and Sell, P. D. (Editors). *Critical Supplement to the Atlas of the British Flora* (1968)

Phillips, W. W. A. *The birds and mammals of Pagham Harbour* (Bognor Regis Natural Science Society, Bognor Regis, 1965)

Ragge, D. R. *Grasshoppers, crickets and cockroaches of the British Isles* (1965)

Royal Society for the Protection of Birds. 'Reserve News' in many issues of *Birds* (1966–71)

Salisbury, E. *Downs and dunes* (1952)

Sandwich Bay Bird Observatory Committee. *Sandwich Bay Bird Observatory* (1970)

Sankey, J. *Chalkland ecology* (1966)

Shorten, M. *Squirrels* (1954)

Smith, M. *British reptiles and amphibia* (1949)

The British Amphibians and Reptiles (1954)

Southern, H. N. (Editor). *The Handbook of British Mammals* (1964)

Surrey Bird Club. *Surrey Bird Report* (1957–70)

Surrey County Council. *Open spaces in Surrey* (1969)

Surrey Naturalists' Trust. *The Surrey Naturalist* (1966–70)

Sussex Naturalists' Trust. Annual Reports (1961–70)

The History of Woods Mill (no date)

Woods Mill Nature Trail (guide for adults)

Woods Mill Junior Nature Trail (guide and 'quiz' for children)

The Sussex Mammal Report (1965–9)

Sussex Ornithological Society. *The Sussex Bird Report* (1962–70)

Tansley, A. G. (revised by M. C. F. Proctor). *Britain's green mantle* (1968)

Thomas, G. 'Gull control and reserve management', *Birds*, 3 no 10 (1971), 246–50

Victoria County Histories of Kent, Surrey and Sussex

Warburg, E. F. *Census catalogue of British mosses* (1963)

Watson, W. *Census catalogue of British lichens* (1953)

White, G. *The natural history of Selborne* (various editions)

Wooldridge, S. W. and Goldring, F. *The Weald* (1966)

Wright, C. J. *A guide to the Pilgrims' Way and North Downs Way* (1971)

Useful addresses

DETAILS OF LOCAL societies without permanent addresses may be had by sending a stamped addressed envelope to the Council for Nature or the appropriate county naturalists' trust. Local public libraries also supply information of this kind.

Botanical Society of the British Isles, c/o Department of Botany, British Museum (Natural History), Cromwell Road, London SW7 5BD.

Council for Nature, Zoological Gardens, Regent's Park, London NW1 4RY.

Countryside Commission, 1 Cambridge Gate, Regent's Park, London NW1 4JY.

Forestry Commission, 25 Savile Row, London W1X 2AY.
  South East Conservancy: The Queen's House, Lyndhurst, Hants.

National Trust, 42 Queen Anne's Gate, London SW1H 9AS.
  South-eastern area office: Polesden Lacey, Dorking, Surrey.

Naturalists' Nature Conservation Trusts:
  Kent: PO Box No 29, Maidstone, Kent.
  Surrey: Juniper Hall Field Centre, Dorking, Surrey.
  Sussex: Woods Mill, Henfield, Sussex.

Nature Conservancy, 19 Belgrave Square, London SW1X 8PY.
  South-east region: Zealds, Church Street, Wyc, Ashford, Kent.

Ramblers' Association, 124 Finchley Road, London NW3.

Royal Society for the Protection of Birds, The Lodge, Sandy, Beds.

Youth Hostels Association, Trevelyan House, St Albans, Herts.

Acknowledgements

THE FACTS USED to supplement my own observations have come from a variety of sources and I am grateful to the many authors, editors and contributors concerned, particularly to those connected with publications listed in the Bibliography.

I am glad to have been able to use photographs taken by staff photographers of the Forestry Commission, a much maligned body which has discharged with distinction the duties given to it by Parliament and which has gone to considerable trouble to provide facilities for naturalists and countrygoers.

I am happy, too, to include examples of the fine work of Leonard and Marjorie Gayton, now resident in Sussex, and of Pamela Harrison who, like her father-in-law and her husband, has earned a place in the annals of ornithology.

I also owe a debt of gratitude to my wife D. who shares my love of the South East and who has acted as driver during all our visits to the region.

This is a good place, too, to thank the many people who helped and encouraged me while serving with the RAF in Kent and later while teaching at the Grammar School at Shoreham on the Sussex coast and at Woolpit School at Ewhurst in the Surrey countryside.

I would welcome constructive comment from both resident and visiting naturalists.

Finally I would thank all those who are on the side of Nature and natural beauty in the South East and elsewhere.

S. A. MANNING

Index

Most of the scientific names of species mentioned are given under their appropriate chapters, and can be found by looking up the page numbers against the English names below. Bird species not listed below will be found in the alphabetical index of birds, pages 182–98. Italic figures indicate illustrations.